MW00330158

BEFORE TIME BEGAN

BEFORE TIME BEGAN

BEGAN

The Big Bang and the Emerging Universe

Helmut Satz

OXFORD
UNIVERSITY PRESS

OXFORD
UNIVERSITY PRESS

Great Clarendon Street, Oxford, OX2 6DP,
United Kingdom

Oxford University Press is a department of the University of Oxford.
It furthers the University's objective of excellence in research, scholarship,
and education by publishing worldwide. Oxford is a registered trade mark of
Oxford University Press in the UK and in certain other countries

Published in the United States of America by Oxford University Press
198 Madison Avenue, New York, NY 10016, United States of America

British Library Cataloguing in Publication Data

Data available

Library of Congress Control Number: 2017943713

ISBN 978–0–19–879242–0 (hbk.)

Printed and bound by
CPI Group (UK) Ltd, Croydon, CR0 4YY

Quid faciebat deus, antequam faceret caelum et terram?
What did God do before he made heaven and earth?

AUGUSTINUS (AD 354–430)
CONFESSIONES II/12

Preface

Ever since humans began to think about the world in which they lived, they have asked themselves how this world came to be, why it is as it is, how it was before, how it will be in the future, and what role we will continue to play in it. All human civilizations try to give answers to these questions. Wherever we look, we see happenings which have a beginning and an end, just as our lives do. Did our universe have a beginning, and will it have an end? When we stand at night beneath the immense starlit sky, it seems natural to ask how all that was formed, how large it is, and what will become of it. We are such a tiny part of something so great, but still we can ask such questions, and perhaps it is that which makes us special.

The search for answers to these questions has led to religions, to epics, to a wealth of philosophies, and eventually, of course, to natural science. In science, cosmology has developed as the relevant branch of research, and what we want to say here is largely based on studies in its framework. Nevertheless, even if different answers to the basic questions have in the course of time seemed to contradict each other, it appears that these differences were rather superficial, and that the essentials are more in accord than initially thought. And we find today that cosmology is entering regions which for many physicists are no longer science, since they are no longer experimentally explorable. Nevertheless, they remain challenges to human thinking.

In this book I want to describe the essential steps in the formation and evolution of our universe, as they have proceeded according to the presently prevalent ideas in cosmology and physics. We will see that much is indeed in accord with previous, less scientific views. Already more than two thousand years ago, much of what today is considered as the latest scientific result was proposed simply as the outcome of logical thinking. What has been added from the time of Galilei on is certainly the insistence that all conclusions must be confirmed by experiment. It is only that which turns metaphysics into physics. Nevertheless, we note that in recent times more and more interesting ideas have been pursued, from a multiverse and parallel universes to wormholes through space and time, even though in our present world these concepts are not accessible to experiment. The world of the imaginable remains much larger than that of the testable, and so in the future, even in natural science, concepts and ideas may well survive without presently having a chance to be verified experimentally.

The basic questions we want to address here fall into three areas:

- How and out of what was our universe formed, our world in time and space? What was before, and what will happen afterwards?
- What are the basic building blocks of matter in our present world, and what forces determine their binding?
- How could a uniform, structureless, primeval world lead to the present multitude of forms and structures?

The answers we have today to these questions are, as already indicated, still of a somewhat speculative nature, and they are certainly not accepted by all. But I believe that they are interesting enough to pursue them further. That is the aim of this book.

Less than thirty years ago, the first of these questions was still ruled out as politically incorrect. The beginning was the Big Bang, and "before" made no sense; there was no "before." Today, many cosmologists and physicists picture the birth of our universe as a rapidly expanding bubble in a hot primordial world, one bubble among many others. We are witnessing today a second Copernican revolution: neither our solar system nor our galaxy nor our universe are the end of all things. Beyond our world there are innumerable others, similar or not similar to ours, worlds we can never reach but which should nevertheless exist. These views make the Big Bang a physical process and not a singular occurrence—it is not a unique event; there were and there will continue to be others like it, and it can also lead to an end.

The question about the building blocks of our present world and of its predecessors in earlier stages of development can today be answered in a more extensive way than ever before, thanks to progress in particle physics. The dream of a final theory, a theory in which electromagnetic, strong, and weak nuclear forces are unified into a single interaction, is still not fulfilled, but has become more conceivable. Such a theory of grand unification must describe a primordial world of great symmetry, in which all constituents are treated equally. The cooling of the universe then leads to the breaking of symmetries and thus to the different interactions. The role of gravity in such a picture still remains enigmatic, however.

The development of the complexity of our world has also led to different and perhaps even contradictory ideas. The crucial point here is the famous second law of thermodynamics, insisting on a development toward increasing entropy, and thus giving a direction to time. Does that mean that our universe should become more and more disordered and thus head toward an end without form and structure? Here two ways out have been noted. First, the increasing expansion of the universe can in the long run prevent thermal equilibrium from ever being reached. Second, the role of gravitation as the dominant force implies that in a cooling gas an equidistribution of matter is not the stable form of the medium. As long as gravity rules, a world of galaxies and radiation in empty space is thermodynamically preferable to a uniform gas of particles.

So, for all three questions, much has happened in the past 30 years, and our thinking has been thoroughly modified. Two basic approaches of physics—reduction (What are the smallest possible building blocks of matter?) and extension (Earth, planets, solar system, galaxy, supergalaxy, universe)—seem to have reached their limits, through quarks and the multiverse, respectively. On the other hand, the new concept of emergence has appeared on the scene. We distinguish today fundamental observables and forces (charges, atomic binding) from emergent observables and forces (temperature, pressure) that arise from the collective effort of many constituents. It seems meaningful to present these new views of the world to a general, nonspecialist readership, and that is what I want to do here. In doing so, I want to concentrate more on the novel concepts, notions, and ideas that have come up, and less on details about who did what, when, and why. It could well be that the new conceptual developments will eventually have profound consequences in other areas of human thought—it would not be the first time for physics. In any case, I want to show to all those interested that we are today witnessing the emergence of new ideas in natural science—ideas that so far have had a much greater impact on our thinking than they have on our technology. Whether they will ever result in advances of practical use to mankind—that remains to be seen. But they have already fundamentally changed our view of the world in which we live. *The* universe was reduced to *one* of the universes in the multiverse, one among many in a primordial world. We have become a smaller part of a greater whole.

When writing this book and looking at the various approaches, it became clear to me that in modern cosmology there still remain different, even contradictory views of the beginning of our universe. My guiding line therefore was the approach of the proverbial American baseball umpire who noted, "I calls them the way I sees them." Others may have different views, and they may be as valid as the one presented here.

A slightly earlier version of this book has appeared in German, entitled *Kosmische Dämmerung (Cosmic Dawn)*, from C. H. Beck Publishers, Munich. I am very grateful to various colleagues here in Bielefeld for discussions on many aspects of the subject; Frithjof Karsch looked through all the entire earlier version and helped improve it greatly. Particular thanks also go to Paolo Castorina from the University of Catania, Italy, who was always ready to join me enthusiastically even in the pursuit of topics that traditional physics considered as devious. Last but certainly not least, most sincere thanks go to my wife, who once more found herself in the role of wholeheartedly supporting the musician without hearing the music.

Helmut Satz
Bielefeld
November 2016

Contents

1

Before the Big Bang

In the beginning, God created the heaven and the earth.

THE BIBLE, GENESIS I.I

Before the Big Bang, there was no time and hence also no beginning. In six days, so the Bible says, did God create the world. Although omnipotent, he did not do it with one mighty stroke, but instead spread creation over six days. Light appeared on the first day, land and water on the second, their separation occurred on the third, and so on. The theologians of earlier times called the creation process the *Hexameron*, the work of six days. Why did creation take so long, and why did it pass through different stages? It would be too simple to blame that on a human-like God who became tired in the evening—after all, he had just created morning and evening—or who needed more time for further planning. Instead, it seems more natural for us to imagine that something as great and as complex as our present world could simply not be formed at once, that even God could only have it emerge from something simpler "in the course of time."

That in fact shows the essential result of a sequence of successive steps of creation. Had the world been made with one stroke, it would have been timeless. Only the successive occurrence of different phases of creation introduced the idea of time, a course of events and intervals between them, defining a scale. The progression of happenings also defined a *direction* of time, and the first event in this progression then became the beginning.

In a similar way, space first appeared. Only the separation of heaven and earth, of land and water, defined a space in which certain

Hildegard von Bingen (1098–1179)
The Hexameron

things are here, others are there; some on top and others on the bottom. Thus, the stage for the coming play first had to be created, and so, in today's thinking, the opening line of Genesis could well be "In the beginning, God created space and time."

This makes it natural to ask what was before, how and from what space and time were formed. What did God do *before* he created heaven and earth, or as we read it, before he made space and time? This seemingly heretical question had in fact been put forth in early Christian theology. Augustine Bishop of Hippo in North Africa and one of the founding fathers of the Christian church, around AD 400 comments quite critically on the answer that before heaven and earth, God made hell for those who ask such questions. Augustine himself comes to the conclusion that before creation God did nothing; the making of heaven and earth was also the beginning of time and hence the beginning of all actions.

Similarly, today's cosmology is based on a Big Bang as the beginning of our universe. Distant galaxies recede from us at ever-increasing speed, and if we let this film run backward, we come to a point at which it all started, a point that the Belgian priest and physicist Georges Lemaitre proposed as the beginning of our world. And, just as the work of God before creation was ruled out, it was for a long time not considered meaningful to ask about the state of the world *before* the Big Bang, or how the Bang came to take place. Not so long ago, even Stephen Hawking, famous for his *Short History of Time*, noted that these questions were like asking what lies north of the North Pole.

The past thirty years, however, have seen a change of paradigms, leading to a new view of the world, one in which the question of how and from what our universe appeared makes sense after all. We do have a frame today, still speculative and certainly not accepted by all scientists, that permits us to discuss in general terms the formation of our universe. We can now relax and wonder what God did *before* he made our world.

Our view of the world began with the Earth as the center of everything. This geocentric scheme was then replaced by a heliocentric one, with the sun at the center and the Earth only one of the planets circling around it. Still later, it was found that the sun, our sun, is only one of the millions of stars making up the galaxy of the Milky Way. And today we know that there exist millions of similar such galaxies, spread out in space and moving away ever farther through the expansion of space. Our own world thus became an ever-smaller part of an ever-growing universe. The Big Bang defined the beginning of this immense universe and led to the idea that this is all there is, *the* universe. Four hundred years ago, Giordano Bruno could imagine an endless succession of solar systems; today's cosmologists can imagine an endless number of universes, similar to ours or different, with the same or with other laws of nature. Ours is but one among many universes in this infinity, in what is now called

The multiverse.

How, when, and from what could all these universes have appeared and continue to appear? What are the essential features of the multiverse?

The world before our time began is the primordial world. At that time, there was no time in our sense. The course of time requires a sequence of different events, allowing us to identify before and after. The primordial world has no beginning and no end, no before or after, no earlier or later, no here or there, no top or bottom, no large or small, no form or structure. Two thousand years ago, the *Rigveda*, the Indian epic of creation, said that

> *At that time there was neither being nor not being*
> *Only darkness hidden by darkness,*
> *And invisibly all moved about.*

As an accidental fluctuation in this primordial world, like a bubble in hot lava, our universe appeared, with its space and its time. The *Rigveda* says

> *The One was born from the power of Fire,*

but in the same way many other bubbles, many other universes appeared as well, and they continue to appear. It is not easy to envision such a primordial world and the eternal creation it provides. Space and time are such basic features of our world that it is difficult to imagine a world without them. We can try, however, by starting with one design, and then keep on correcting that as we go on.

Let us imagine an immense container filled with water, an ocean at fixed temperature and without any external interference. We dive deep into this water, far from all confining boundaries. Here we find only uniform water, yesterday, today, tomorrow; time is meaningless, and the idea of space as well, since a shift in position, no matter in which direction, does not lead to any change in the world around us. And if we picture such an ocean placed in interstellar space, without any effects of gravitation, there would also not be any up or down.

As soon as the chosen temperature approaches the boiling point, small bubbles of steam form, regions of lower density than that of the surrounding water. In a terrestrial environment these bubbles rise, escape into the air above the surface of the water and continue to expand. That would be a very simple picture with which we could start to describe the formation of a universe: the hot water would be the primordial medium, the multiverse, and the bubbles would later on become universes of some kind.

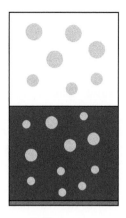

Boiling water

But we can make this even more interesting. If the water is very pure and the walls of the vessel very smooth, the temperature can be increased beyond the boiling point without any effect, in practice up to some 110 degrees centigrade; physicists call this *superheating*. The water is now in a metastable state: any small disturbance will cause an explosion, in which a big bubble of steam escapes from the medium. Once the bubble is out of the water, its density will decrease more and more; now there is a time with a direction. Before, in the water, that was not the case: water is water, today or tomorrow; the question of time does not arise. Time only comes into play once the bubble has escaped, once a chain of events appears. The water molecules now separate more and more and travel into ever more distant regions of space. The Big Bang, according to today's experts, was a somewhat similar process.

Let's stay with this experiment for a moment. Under fixed conditions, water is in a *normal state*: under atmospheric pressure at sea level, it is ice below 0 degrees centigrade, liquid from 0 to 100 degrees, and steam above 100. The transitions from one normal state to another, like melting or evaporation, are called *phase transitions*. But we just saw that if we are really careful, we can heat water some ten degrees above the boiling point without evaporation setting in. It is still liquid under conditions when it should really be steam. It is now, using physics terminology, in a *false* normal or ground state, and any disturbance will cause it to flip to the right one. Superheated water thus was artificially brought into an unstable state of too high an energy, and in the transition from wrong to right it will liberate this energy: everything flies apart, the bubble explodes. Incidentally, a similar phenomenon occurs if we cool the water carefully below the freezing point; any disturbance then causes sudden ice formation. Supercooled rain coats everything it hits with a layer of ice.

In our daily world, we encounter many other such cases. A well-known instance is the ball on the hill; here as well, any slight disturbance will cause it to roll down. On top, in the wrong, unstable state, the ball has a higher gravitational potential energy than it would have on the bottom. This higher potential energy then turns into kinetic energy of motion when the ball rolls down.

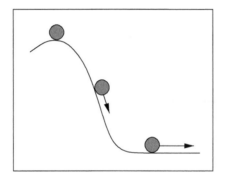

The ball on the hill

What we have said so far points us in the right direction, but it is still too much based on our familiar terrestrial world. The universe has features that cannot really be modeled by water. In particular, we know today that the universe is not static; it is undergoing continuous spatial expansion. Our terrestrial world provides a static stage for the ongoing events, but in the cosmos things look different. Distant galaxies are becoming ever more distant and eventually even disappear; on large scales, everything is expanding, and this has serious consequences.

Imagine you are on one side of a large room and now walk toward a door on the opposite side, with your normal walking speed of about a meter a second. If the room expands by more than a meter a second as you are walking, you will never reach the other side. In fact, even though you walk and walk, the door becomes ever more distant.

A similar fate is in store for an ant trying to crawl on an expanding balloon from the equator to the North Pole. If the balloon is blown up rapidly enough, the poor ant will never reach its goal; the pole will move farther and farther away.

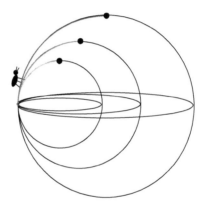

The ant on the balloon

We conclude from these examples that a sequence of events can be seriously modified by spatial expansion. The crucial question is evidently how the rate of space expansion compares to the time scale of the process we are looking at.

Before we return to the expansion of the cosmos, we have to address another essential problem: there has to be some agent causing this expansion, since gravitation provides an attractive force between all stellar objects and, as we know from Einstein, even between clusters of normal energy. The mysterious something that not only compensates this attraction, but even overcomes it to cause expansion could be called

Space energy,

but in common usage it is now generally referred to as *dark energy*. From the point of the view of cosmology, empty space is therefore not really empty; it is filled uniformly with an invisible energy of very low density, driving it continuously to further expansion. Normally, such an expansion would cause the energy density to drop; here, however, that is not the case: it remains constant in space and time, having everywhere the same value, usually denoted by Λ. Such a concept was first introduced by Einstein, who called it the *cosmological constant*. At first sight, such a constancy in space and time does not seem possible: if the total volume grows and the energy density remains constant, the total energy of the universe is continuously increasing. Where does that energy come from? Isn't that a violation of energy conservation?

The solution to this puzzle is found in the description of gravity through space deformation, as it is formulated in the general theory of relativity. We saw a first instance of this in the case of the ball on the hill. The kinetic energy of the ball is zero on top of the hill, but clearly not zero when the ball hits the bottom. How is this kinetic energy created? The topology of hill and valley leads to a difference in gravity, and that difference in gravitational potential energy provides the needed kinetic energy. In a similar way, energy and space are related in the cosmological world: the increase of space energy is paid for by an increased deformation of space. The bigger the effective mass of the universe becomes, the bigger the negative potential energy of the resulting gravitation. Only in this way was it possible to extract a universe as "the ultimate free lunch" out of the primordial world: what was gained in mass was paid for by the resulting negative potential energy of gravitation.

The present density of dark energy in our universe is extremely low, as already mentioned: its mass equivalent in a volume of the size of the Earth is about one-thousandth of a gram. In stellar dimensions, this does not lead to noticeable gravity effects, and it also does not affect physical processes. Our planetary system and even the Milky Way remain unchanged; gravity still wins. Only on cosmic scales—the astrophysicists speak of intergalactic scales—do we find an effect: since the density is constant over the entire universe, the sheer total amount of dark energy can now overcome gravity and cause a repulsion. In our solar system, the total amount of dark energy is negligible, but for the total observable universe it adds up to three-quarters of the total energy, and because of the expansion, that fraction is continuously increasing.

According to present cosmology, the dark energy in our universe *after* the Big Bang plays the role of the steam in the escaping bubble. *Before* the bubble escaped, it contained a primordial medium of immensely higher density. This created a correspondingly higher rate of expansion than the present one; the primordial world expanded and continues to expand much more rapidly than our universe. The primordial medium is the counterpart of the superheated water; it is in a wrong normal state, and any small local perturbation can create a bubble in which the transition to the correct normal state sets in. The high density of the primordial medium lets the bubble initially expand with the dramatic primordial rate, until the wrong normal state has been converted into the right, and the dark energy density has dropped from its primordial value to its present one. From this point on, the expansion rate also decreased correspondingly. The energy difference between the high false primordial density and the lower stable one today is liberated in the transition, and eventually gives rise to all the matter in our present universe. All this makes the Big Bang a "normal" physical process; nevertheless, for us it is the beginning of what we consider as space and time.

The origin of our universe as a bubble of primordial medium is, as indicated earlier, a new view of the world, only some thirty years old; a major role in its formulation was played by the American physicist Alan Guth and the Russian-American physicist Andrei Linde. Until then, our universe was considered as ultimate and final, not as part of something bigger, and for that reason, the question "What was before the Big Bang?" had been ruled out.

Alan Guth
Photo: Jenny Guth

Andrei Linde
Photo: Linda Cicero, Stanford
News Service

Nevertheless, it had been found that different observations—we will soon come back to them—required that very shortly after the Big Bang, the universe must have gone through a phase of extremely rapid expansion, an "inflation." Attempts to justify such an inflation then led to the new primordial scenario presented here. This scenario has other unavoidable consequences as well. Our universe emerged from a bubble of the primordial medium; however, that medium continues to exist and expand further, so that new bubbles continue to appear forever. Hence, the primordial medium is a *multiverse*, now and forever giving rise to new universes; ours is only one of all these uncounted worlds. In the following picture, we have for simplicity made the new universes round; in reality, they can be arbitrarily irregular.

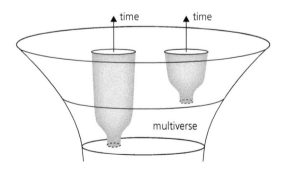

Two universes appear out of the primordial multiverse

Because the other universes remain forever inaccessible to us, we will never know if the same laws of nature hold there, or if they might even hold intelligent life. For us, they are beyond any form of investigation, and for this reason many physicists consider the idea of a multiverse to be heretical, more metaphysics than physics. One of the most eloquent critics is the American theorist Paul Steinhardt, even though he initially helped to formulate the idea of a multiverse. On the other hand, a single unique universe also leads to problems such as inflation, which are difficult to comprehend in a conventional scheme and seem to require further explanation.

On the largest cosmic scales (the mentioned intergalactic scales), our own universe is, as far as we can tell, quite uniform: it is homogeneous as well as isotropic, i.e., it is of the same form in different regions and in all directions. This feature is generally referred to as the *cosmological principle*. Of course, it only holds for an average over immense regions; in our solar system and on to the Milky Way, our world is certainly not uniform. It only becomes so once we average over cosmic distances; different spatial regions and directions then become indistinguishable. While that applies to space, it is certainly not the case for time: ever since the Big Bang, our world has been steadily and dramatically changing; there is a direction, an arrow of time that leads from a hot primeval gas to a cosmos of galaxies.

Things look different in the case of the multiverse. Here as well, one cannot distinguish sufficiently large spatial regions, but they are far from uniform. In the false normal state of the primordial medium, one finds everywhere expanding bubbles undergoing transitions to the right normal state, creating future universes. And the multiverse itself continues to expand dramatically; it is a bit like a superheated soup, from which bubbles are continuously escaping. This behavior does not change with time; it goes on forever, so that for the multiverse, space and time are now on equal footing: the bubbles of the future universes appear here and there, bigger or smaller, now or later. The superheated multiverse itself undergoes an ever-increasing expansion as well.

A medium showing such irregularities is today called *fractal*. This concept was introduced by the French mathematician Benoit Mandelbrot; it refers to complex structures formed through the repetitions of a given form at different scales; it is also designated as *self-similar*. In the case of mathematics, one considers such structures in space, but in the case of the multiverse, they appear in both time and space: bubbles appear here and there, now and later, larger and smaller. For illustration, we show in the following figure one such structure; it is a triangle formed through the repetition of black and white triangles of different sizes and named after its inventor, the Polish mathematician Wacław Sierpinski.

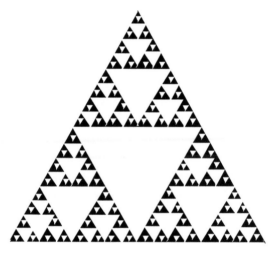

The Sierpinski triangle

In the case of the multiverse, one has to picture the whole structure to be in continuous expansion, both in space and in time. And even then the picture is not quite correct: if we take the white triangles to be new universes, then, as far as we know today, the black primordial world expands much faster than the white, leading in turn to more new universes.

In the end, that means our entire universe, from the Big Bang until today, is only one of an ever-growing number of universes, all created in the same way. Copernicus had abolished the idea that the Earth was the center of the world and thus taken away our special cosmic status. Today's cosmology takes such a status away even from our universe. It is only one of the many little triangles; we don't know and never will know what is happening in the others. Such a scenario of course leads to various questions. Our present world is possible only because our constants of nature have the values they have. If, for example, electrons were much heavier or protons could easily decay, a world such as ours could not exist. So why do they just have the values required for our existence? One proposal—not really all that satisfactory—is that the values of the constants of nature in the different universes of the multiverse have randomly distributed values, and ours just happen to have the ones they have.

Up to now, we have considered the formation of our universe "from the outside"; let's now return to our own world. In the twenties of the last century, Edwin Hubble made astronomical observations at the Mount Wilson Observatory in California; they showed that distant galaxies are moving away from us, and that their recession speed increases with their distance from us. The reason for this, as already mentioned, is the expansion of entire universe, driven by what remained of the dark space energy in its present normal state. If we now have the film run backward, we come to a beginning, to

The Big Bang.

Subsequent measurements indicated that this must have been about fourteen billion years ago. In 1964, the American physicists Arno Penzias and Robert Wilson found what has become the most essential confirmation of the Big Bang theory: the *cosmic microwave background radiation* as the surviving light from the Big Bang.

Edwin Hubble (1889–1953)

Part of the energy liberated in the transition from wrong to right normal state eventually turned into photons, particles of light. In the early stages of development, these photons suffered interactions with all other charged particles that were created. But sometime later on—we shall return to this in detail—many of the charged particles annihilated each other, bringing still more photons into the game, and they all were in interaction with the remaining charges. In the long run, however, those remaining charges combined, forming neutral atoms from positive protons and negative electrons. For the photons, that was the ultimate liberation: they were now decoupled from matter and since then, they could travel on freely. This decoupling or last scattering occurred some 380,000 years after the Big Bang. The photons present at that time have ever since been on their way, flying freely through space. Space, however, kept expanding, and, as a result, the wavelength of the photons increased correspondingly. Because the wavelength determines the temperature of the photon gas, the implication is that the temperature of the cosmic background radiation dropped from about 3000 degrees Kelvin at the decoupling time to about 3 degrees today. It was just this 3 degree radiation that Penzias and Wilson found. By now, there are many ever more precise measurements of that radiation, in all regions and directions of the sky. And wherever one measures, one finds the same 2.72548 ± 0.00057 degrees Kelvin, up to the fourth decimal position. The universal remaining light of the Big Bang is thus known very well and it is with a high degree of precision the same everywhere. That, as we shall see shortly, creates a serious problem.

At the time of decoupling, a small region of the universe at the eastern rim of the sky was many light years distant from one at the western rim. Between the two regions, no communication was possible; each was outside the other's *causal horizon*. So how could they synchronize their radiation temperatures with such precision? How could the conductors of two orchestras in absolutely separate concert halls, one in London and one in Sydney, manage to start the identical music at precisely the same instant? This puzzle, this horizon problem, has bothered cosmologists for decades; today it is thought that it is solved by the idea of

Inflation.

At the beginning, there was a small bubble of hot, dense primordial matter in thermal equilibrium, like a gas at fixed temperature and pressure. Then, suddenly, the inflation set in, the transition from wrong to right normal state. In an unbelievably short time, space expanded by an immense factor, thus creating regions that could no longer communicate. And in this expansion, even the slightest irregularities were largely smoothed out. Such a process lets us understand how causally disconnected systems can still have the same temperature: they were together in the same pot and under the same conditions, until the explosion-like inflation pulled them light years apart. What is separate today was once together and could communicate. The relevant scales are illustrated in the following picture. It shows that the inflation process had lengths increase by a factor of 10^{26} in a time interval of only 10^{-34} seconds. This means that the space of the primordial bubble expanded at a rate that surpasses the velocity of light by a factor of 10^{50}. So it was indeed a very special process, difficult to accommodate in the previous views of physics.

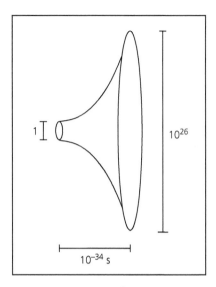

Cosmic inflation

The new view of the origin of the universe thus solves two puzzles at the same time. We see how our universe appeared as a bubble in the primordial world, and how the inflation of that bubble allows causally separate regions to exhibit the same conditions. The cosmic background radiation thus contains more information than was thought at first. It provides our only tool to look for something remaining from the flash of the Big Bang, of the transition from wrong to right normal state. And the universal temperature value of the radiation implies that the transition must have been accompanied by a dramatic spatial expansion, by cosmic inflation. That brings us to the next question: how could the transition terminate, in these first instants of the Big Bang, with the formation of something like matter?

2

The First Particles

I call them elementary particles,
because they were the first and everything is made of them.

LUCRETIUS, *THE NATURE OF THE UNIVERSE*, ABOUT 55 BC

The first particles appeared when the metastable medium of the bubble in the primordial multiverse had fallen from its hot but false ground state into the much cooler right one. This drop suddenly liberated a lot of energy. The transition—which in our picture of superheated water corresponds to the explosive escape of an expanding bubble of steam—leads here as well to a very violent explosion, the "inflation" of the cosmologists. In an incredibly short time, some 10^{-34} seconds, a spatial expansion took place by a factor of 10^{26} or more. At the end of this process, the system found itself abruptly in the correct ground state, together with an excess of energy determined by the energy difference between false and right ground state. There is no comparable "moment" in the physics of our world, and no comparable expansion rate. The whole process, though a phenomenon within the scope of physics, was an absolutely unique event in the history of our universe, its birth: the Big Bang.

Let's look at the event in a little more detail. A density fluctuation occurs in the boiling primordial medium of the multiverse; quantum mechanics tells us that such fluctuations randomly appear again and again, here and there, with a certain probability. Now, it is sufficient if this disturbance creates a bubble in which the system is driven from the false to the right ground state. As long as the energy density in the bubble corresponds largely to the false ground state, the bubble continues to expand with the dramatic rate of the multiverse. Then, suddenly, the transition, the crash into the energetically much lower right ground state occurs, and the expansion is essentially stopped: inflation is over. Our universe is born, and from now on what becomes relevant is what so far was called the Big Bang cosmology: the development of the universe *after* the Big Bang.

The crash into the right ground state means, as mentioned, that the energy difference between false and right ground states is suddenly thrown into space. So what happens to this immediately available excess energy? The example of the ball on the hill showed that the potential energy present on top was transformed into kinetic energy at the bottom. And this could be employed in different ways. If the ball were made of glass, the impact could lead to an "explosion" into many fragments. In the case of the primordial bubble, the liberated energy provided enough fragments to make up the entire world to come.

To understand that, we first have to consider what happens when energy is deposited in empty space. What actually is *empty* space? Here we leave aside the dark or space energy mentioned in the last chapter; it shapes space, makes it expand, but it does not provide its contents. From the work of the British physicist P.A.M. Dirac, we know that empty space is empty only at first sight: it is in fact actually a sea of "submerged" virtual particles, which so far could not come into reality because they lacked the necessary energy. Empty space is like a silent lake, under whose surface all sorts of particles only wait to get the energy for a jump upwards, into reality. What kind of fish are swimming there? What are particles, and what species of particles exist?

In today's world, so particle physics tells us, there exists a multitude of different species of particles, interacting with a number of different forces. But this diversity has developed only in the course of time after the Big Bang, just as the different forms of living beings have arisen through evolution from a simpler primeval form. It is therefore not surprising that physicists would like to find one such basic primeval form, from which all others emerged. This ambition has led to various attempts, but so far not to a definite solution; we return to that shortly. Here, however, we note that shortly after the Big Bang two basic forms had already arisen through the input of the available energy:

Matter particles and force particles.

The former are the basic building blocks of matter; everything is made of them. The latter convey the interaction between these building blocks. They are the cement holding the blocks together, and they are also emitted or interchanged when two matter particles collide, as a signal of such an event.

If we divide matter into ever-smaller constituents, we expect to find a limit: the smallest building blocks of matter. Such a limit, however, only makes sense if no further reduction is possible—if there is an end to divisibility. Because of this, many early attempts failed. The atoms of the chemists were found to consist of nuclei and electrons; the nuclei of the physicists are made up of protons and neutrons; and these, we believe today, in turn are bound states of quarks. The Roman philosopher Lucretius, author of the celebrated early treatise on *The Nature of the Universe* (source of the quote at the beginning of this chapter), had therefore insisted over two thousand years ago that the truly ultimate constituents of matter could never have an independent existence, but that they could appear only as a part of something bigger. If they could exist individually, we could continue to ask what they were made of. Therefore, he concluded, the ultimate building block *can never exist by itself, but only as a primordial part of a larger body, from which no force can tear it loose.* That would provide a logical end to divisibility. And today, quarks indeed fulfill this requirement of Lucretius: they are, in the terminology of present physics, subject to "quark confinement"—one quark can never exist on its own. Nucleons consist of three coupled quarks, and no force can ever split them. That was shown by numerous experiments, and our basic theory of their interactions ("quantum chromodynamics") contains such unbreakable binding as an essential prediction.

But right after the Big Bang, that was not a problem. There was no empty space, no *nothing* of any kind. The liberated energy was so huge that it led to an immensely dense cloud of buzzing particles, of quarks. No quark had ever to fear being left alone; everywhere, in the most immediate neighborhood, there was a multitude of other quarks. That leads to a rather amusing situation. As part of a nucleon, a quark is forever bound to its two partners; no force can ever tear them apart. In contrast, in the dense primeval crowd, any quark is completely free: it can go wherever it wants to; never is there a vacuum; a quark always finds more than enough of the companions required by quark confinement. It can move over arbitrarily large distances, always accompanied by new, ever-changing partners. Such a primeval medium is today denoted as "quark plasma," and different projects of large-scale experimental research are presently attempting to produce such a plasma in the laboratory.

But these primeval building blocks possess another crucial property. Why does a nucleon consist of three quarks and not of many more? Why does the size of a nucleus increase with the number of nucleons it contains? Apparently, it is not possible to accommodate arbitrarily many particles in a given volume. Each matter particle seems to insist on having its own spatial volume, no matter how small—but a finite space just for itself. The sum of all these volumes with the particles living in them then gives us matter. Indeed, we have in today's physics such "territorial" particles, which do not allow another identical particle in its space: electrons, nucleons, and—as constituents of the nucleons—also the quarks. They all are subject to the celebrated *exclusion principle* formulated in 1925 by Wolfgang Pauli: it is not possible that two completely identical particles of this kind can exist at exactly the same point of space.

An immediate consequence of this principle is that atomic nuclei must grow in size with increasing atomic weight, that is, with an increase in the number of nucleons they contain. A gold nucleus contains 200 nucleons, a helium nucleus only four. Since each nucleon insists on its own space, the gold nucleus must be correspondingly larger than that of helium.

Wolfgang Pauli (1900–1958)

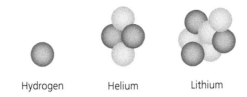

Hydrogen   Helium   Lithium

Nuclei growing in size with the number of nucleons

The exclusion principle, like many concepts in physics, is based on symmetry arguments. One finds that in the underlying quantum theory an interchange of two identical particles at the same place does not lead to exactly the same state, but instead to its mirror image. Now, a given state cannot be the same as its mirror image—in the mirror, your right arm becomes your left arm! But the interchange of two identical particles leaves the world unchanged, and therefore the presence of identical particles at the same place is not allowed. Particles of this kind, with such a symmetry property, are today called *fermions*, named after the famous Italian physicist Enrico Fermi, who developed the basic theory for their physics. Fermi was probably one of the last physicists who made seminal contributions to experimental as well as to theoretical physics; in the 1940s, he played an essential role in the development of the first nuclear reactor.

Matter thus consists in principle of fermions, but they are interacting with each other. Since Einstein, we know that there is no instantaneous interaction at a distance; if two electrons interact, one has to send a signal to the other, and the transmission of that signal takes place with a finite velocity, the speed of light. One electron sends out a messenger, a light particle or photon, which reaches the other after a certain time and passes on the transmitted information.

Enrico Fermi (1901–1954)

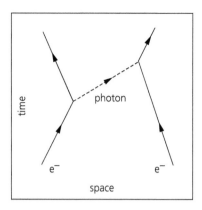

Electron interaction through photon exchange

Thus the second basic form of particles are the force particles mentioned earlier, which provide the interchange of information between fermions, and which also act as cement in the construction of matter. They are not a countable structural part of matter, and so they don't require any proper space. The number of force particles in a given volume is not restricted, and if one carries out an interchange of two such particles, one obtains again the same state. Particles obeying this form of symmetry are called *bosons*, named after Bengali physicist Satyendra Nath Bose, who as a young researcher in 1927 had determined this essential property of photons and sent his results to Einstein. Einstein immediately recognized the importance of Bose's results, translated them into German and made sure that they were published in Germany under Bose's name. Besides the photons, we know today the vector bosons of the weak nuclear force as related force particles; that force makes possible the radioactive decay of heavy nuclei, such as uranium. In the realm of strong nuclear interactions,

Satyendra Nath Bose (1894–1974)

for the interaction of the quarks, the corresponding role is played by the *gluons*; they make possible the binding of quarks to protons or neutrons, to which we return soon.

The basic particle forms for matter and force thus are.

Fermions and bosons.

Fermions are the building blocks of matter; bosons transmit and announce their interaction. At this point, we want to return briefly to the idea of having at the beginning just one primeval form of particle. Is it not possible that fermions and bosons are the descendants of a single particle species? The two present forms are distinguished by their different behavior under the interchange of two particles at the same place: fermions turn into a mirror image, bosons into the same state. The constituents of a greater, unified theory would have to allow both, and so the attempts to formulate it are referred to as *supersymmetry*, SUSY for short. If such a theory was valid in the very early universe, then it must have been possible at that time to convert fermions into bosons and vice versa. Subsequently, there must have been a transition in which this possibility was destroyed. How something like that can happen we will discuss in more detail in the chapter on transitions. So far, there are various attempts to construct such a supersymmetric theory, but they have remained attempts. However, one result of the considerations up to now is that if the primeval world was indeed supersymmetric, then there should remain even today, after the breaking of the symmetry, for each particle a very heavy "supersymmetric partner." For the electron, there would be a bosonic S-electron; for the bosonic photon, a fermionic S-photon or photino, and so on. One has not only given the missing partners names; they were also searched for intensely in high-energy theory and in high-energy collisions, and the search goes on. So far with no success.

As mentioned, we have today various species of matter particles, interacting through different forces; hence there are also various species of force particles. But in the early stages, just after the Big Bang, the world apparently contained only two kinds: primeval fermions with a universal interaction, and primeval bosons, which mediated that interaction. Both were massless, and there was no way to identify or distinguish different species of fermions or of bosons. But the information, which later on led to the different species, must have already been present in some latent form. The seemingly identical particles must even at that early time have had properties which then were unimportant, but which later became crucial. As an example of such a situation, consider two heavy balls of equal size, weight, and appearance, one made of stone, the other of iron. As long as we can only weigh them, use only gravity, we cannot distinguish them. But when we bring in a magnet, one will be attracted, the other not.

In the course of time, the primeval fermions would give rise to all the elementary building blocks of the present world, and the primeval bosons would do the same for the force particles. To see which hidden properties primeval particles must have had, let us elaborate a little on what basic properties the ultimate particles can in fact have.

Besides the requirement of indivisibility and that of its own spatial region, the essential aspect of matter particles is that they should have some property that one can enumerate or count. In the case of electricity, we call this property the *charge* of the particle. If we consider a system of two electrons, each one has an electric charge -1, the whole one of -2. It is precisely this charge that allows us to speak of one or two or three particles. Such charges are immensely important, because innumerable experiments have shown that without outside interference, the total charge of a system will not change, that it is a *conserved quantity*. If a system has total electric charge zero, then the addition of energy to the system can lead to the production of a pair, a positive and a negative charge, but never to just one charge. The positive partner of the electron, the positron, is its *antiparticle*. When the two meet, they can annihilate each other and turn into radiation. Besides the electric charge, there must thus also be some kind of matter charge, which is opposite for electron and positron, so that the sum leads to zero. This charge is expected to be conserved as well.

We might note at this point that while the electromagnetic as well as the weak and the strong nuclear forces lead to discrete charges and their conservation, there does not seem to exist a fundamental charge for gravity. This could be an indication that the force of gravity is indeed of a different nature, as has been proposed recently by the Dutch theorist Erik Verlinde, arguing that it is emergent, the result of a collective effect of many particles. This would presumably rule out a theory of everything, in which all the interactions are unified.

Charges have an immediate and fundamental consequence. The primeval fermions were created out of the liberated energy of the false ground state, and the primeval medium had, as far as we know, no charge of any kind. For this reason, the energy thus provided should have led to equally many fermions and antifermions. Such an inherent duality of opposing concepts seems ingrained in many parts of human thinking; it appears to be completely natural: there has to be both plus and minus, left and right, good and bad, heaven and hell, ying and yang. In the beginning, the world was neutral, and on the whole it should remain that way.

The law of conserved charges seems to have suffered one exception, which has created trouble for physicists for a long time: why does our universe consist of matter, containing, as far as we know, no antimatter? We can of course just assume that it started that way from the very beginning, but explanations of the kind "it always was that way" are considered unsatisfactory, unpleasant by most physicists. So one assumes that the bubble from which our universe was created initially contained no particles and thus had zero charge. That means when the provision of energy through the transition to the right ground state allowed various kinds of fish to emerge from the sea of virtual particles and become real, the fish always had to appear in pairs, fish

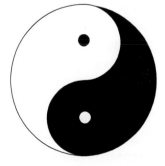

Yin and yang

and antifish. That would conserve all possible charges. One thus assumes that before the Big Bang, there was no matter in our present sense; afterward, matter and antimatter in equal quantities. How did the antimatter disappear, where did it go? Why is the electric neutrality of our present world achieved in such an asymmetric way, with light electrons and heavy protons? We shall return to this issue several times, but we can note a warning even at this stage. The Russian theorist Mikhail Shaposhnikov concluded in a talk some ten years ago that he knew of exactly 44 explanations for the asymmetry, and that he would much appreciate hearing about possible others. That shows how nicely one can say in physics, "We just don't know."

Before we list the fish created in the Big Bang, we have to engage in a little particle zoology: what kind of particles exist today? After what we just said, that means asking what kinds of charges exist in nature. This question has been for quite some years a central topic of particle physics, and so we have to see what that has led to. We have already mentioned electrons, the smallest units of electric charge, and their antiparticles, positrons, that carry the opposite charge. The electron was known for quite a long time, from atomic structure considerations; the positron was predicted in 1928 by Dirac, precisely on the basis of charge conservation, and it was experimentally confirmed shortly afterward by Carl Anderson. Both theorist and experimentalist were awarded the Nobel Prize.

In our present, electrically neutral world, the negative charge of the electrons is in atoms compensated by the positive charge of the nuclei. In contrast to electrons, the nuclei are composite objects, consisting of nucleons—that is, positively charged protons—and neutral neutrons. Nucleons are, as mentioned, in turn bound states of still smaller entities, quarks. The underlying scheme for the binding of quarks is a little more complex, however, again based on the conservation of charges, but now on charges of the strong nuclear force. This requires the existence of two kinds of quarks, denoted by u (up) and d (down) to account for our normal world. For each species, there have to be the corresponding antiquarks, out of which one then constructs antiprotons and antineutrons. These antinucleons can be produced in the laboratory (so they really exist), but otherwise they are hardly encountered in our world.

At this point, the attentive reader might ask what this "hardly" is supposed to mean. In interstellar space, one finds once in a while single unbound nucleons flying around, lonely cosmic wanderers. If two sufficiently energetic such nucleons should happen to collide—and that happens quite rarely—the collision can result in the production of a proton-antiproton pair. The antiproton thus produced can now fly off on its own, until it has the bad luck of colliding with another proton. Then the two annihilate each other and turn into electromagnetic radiation. Such cosmic antiprotons exist—not many, but they can even be observed in terrestrial detectors.

Quarks and antiquarks don't just interact with each other—to form nucleons, antinucleons, or mesons (quark-antiquark bound states)—they also interact with electrons. When such interactions were studied, a further, somewhat spooky particle was encountered: the *neutrino*. It is almost or completely massless. What happened was that free neutrons were found to decay into a proton and an electron. However, in such decays the energy of the proton and that of the electron did not add up to the neutron mass; their sum gave less. This missing energy had been carried off by the unseen neutrino; Pauli invented the process, Fermi the name. The right way to write neutron decay thus is $n \rightarrow p + e^- + $ anti v. Electrons and neutrinos are collectively denoted as *leptons*, from the Greek *leptos* = light. They carry a unit lepton charge, which takes on the opposite value for antileptons. That's why the neutron decay produces an antineutrino: then the overall lepton charge remains zero.

So the essential species of matter particles are

Quarks and leptons.

In both cases, one has the corresponding antiparticles, antiquarks, and antileptons. As mentioned, leptons have a *lepton number*, antileptons the opposite. Quarks also have an additional charge derived from the strong nuclear interaction, the baryon number (from *barys* = heavy in Greek); the name arose because nucleons are so much heavier than electrons and was invented before the quark infrastructure of nucleons was known. Antiquarks have the opposite baryon number, so that in the creation of a quark-antiquark pair, both electric charge and baryon number are conserved. The same holds for leptons and lepton number, in addition to electric charge. The conservation of baryonic as well as leptonic charge today seems to be an absolute law. That, however, cannot have been the case at earlier times: if at the beginning there was matter and antimatter in equal amounts, the corresponding symmetry must have been broken at one stage, in order to produce our universe consisting of nucleons (baryons, not antibaryons) and electrons (leptons, not antileptons).

To obtain a common origin of matter particles, we can imagine that in the very early universe, only one kind of primeval fermion existed, which then later split up into quarks and leptons. This primeval form would be massless—there would be no measure, no scale, and only one kind of interaction, apart from gravity. A theory for such a unified world, a "grand unification theory," or GUT for short, is up to now still more dream than reality, not yet quite reached by much intensive research. Since in such a theory, quarks and leptons would simply be different possible states of a primeval fermion, quark-lepton transitions would now be possible: a quark could turn into a lepton and vice versa. At such an early time, there could have been fluctuations leading to more quarks than antiquarks or to more leptons than antileptons. An excess of quarks and leptons at the time when the GUT era suddenly came to an end, when all bridges between quarks and leptons were broken down— that would provide an explanation for the observation of the present asymmetry between matter and antimatter. The symmetry, valid at the time of the Big Bang, was broken at the end of the GUT period. That made possible our world in its present form: a world containing much matter, but, as far as we know, not a corresponding amount of antimatter.

We have thus found a further important threshold in the evolutionary history of the very early universe. There was a point in time at which quarks and leptons went their separate ways. From now on, in each reaction, the total number of quarks (i.e., quarks minus antiquarks) and the total number of leptons (leptons minus antileptons) had to remain unchanged. Whenever a quark-fish surfaced, it had to be accompanied by an antiquark-fish, and the same held for leptons. Since our present world contains so much matter and so little or no antimatter, the particle-antiparticle symmetry was broken at that time, and so we can identify the end of the GUT era as

The birth of matter.

At this point, the transition to the next era led to a few more quarks than antiquarks, a few more leptons than antileptons. This excess was not dramatic: estimates indicate that for 30 million antiquarks, there were 30 million plus one quarks. But this tiny excess now had to be preserved for all times, and it finally was enough to provide all the matter found today in our universe. Small causes can have big effects; how that took place, we shall see soon.

Up to now, all fermions were massless, quarks as well as leptons. Today, both have masses, but at that time they did not. When, where, and how did these masses appear? What exactly are masses? On one hand, they specify the resistance to forces, as *inertial masses*; on the other hand, they measure the effect of gravity, as *weights.* Are masses perhaps something like a charge for the force of gravity? Is there a smallest mass as the fundamental charge of gravity, like the electric charge in electromagnetism? Recent considerations seem to indicate that this not so. Masses are an *emergent* property, not an *intrinsic* one. They were not always there, but rather were dynamically created, through interaction, in the course of time. How can such a *dynamical mass generation* occur?

How can a massless particle suddenly acquire mass? That is a rather common problem in physics, and so we will first look at it in general. The simplest way to gain mass is shown by a snowball: a small light snowball becomes larger and heavier as we roll it in the snow. It attracts the flakes of the medium and makes them part of the ball.

Something like that can always happen when there is an attractive interaction; a certain amount of the medium is then converted into additional mass of the particle. A well-known case of this effect is found in the polarization of electric charges. A plasma of equally many positive and negative particles, which are otherwise identical, eventually reaches an equilibrium in which the opposing charges just compensate each other—in each local volume, there are as many positive as negative. If we now introduce into this medium a strong negative charge, positive charges will feel attracted to it and compensate the new charge, creating a larger object, which is again neutral. The overall result is thus the formation of a new, larger and heavier "particle"—the introduced negative charge with its cloud of positive charges—and this object, being on the average electrically neutral, can move freely through the plasma. The polarization of the plasma charges around the intruder thus has increased its mass. And this new mass is indeed an inertial mass: if we try to move it, we have to move both the original intruder and the cloud of the clinging charges. Polarization thus has really led to a mass increase, and we will encounter other examples of such a mechanism of mass generation.

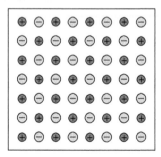 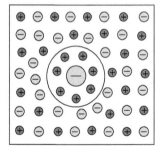

Mass generation through charge polarization

In different and more complex versions, such a mechanism is today quite generally considered as *the* origin of mass. Particles have various inherent labels—electric charge, baryon number, spin, and more. Mass is not such a property, and neither is spatial size. Before the quark mass was created, was there a force of gravity, however weak, between two quarks? That question, though somewhat academic, so far does not really seem to have a final answer.

We have not yet mentioned what the medium is that plays the role of the snow or the plasma, giving rise to the masses of quarks and leptons. We shall return to that issue in the next chapter; here we only note that besides the primeval fermions and bosons, there must exist some further field which acts as the mass-generating snow. But even at this point one aspect becomes evident: if we roll the snowball too fast, it cannot collect further flakes. And the new intruding strong charge in the plasma also needs some time to collect the cloud of opposite charges. In the same way, the primeval universe had to cool down enough after the Big Bang, so that the speed of the primeval fermions was sufficiently reduced for mass generation. When that was the case, the quarks collected clouds of the primeval snow to acquire small intrinsic masses, and so did the leptons; their intrinsic masses were of comparable size, from half to a few MeV, in the units of particle physics.

To have an overview, let's combine all the mentioned species of particles to form a genealogical tree, a family tree. As in all such trees, some members are closer to us than others. Quarks and gluons combined to form nucleons, and these determine our own mass. With the help of photons, nucleons and electrons combined to create atoms, and those are the basic constituents of the present world. At this point, we may wonder if there are particles around today that have been with us ever since the Big Bang. We shall see that that is in fact quite unlikely. All particle forms have passed through various stages of development, in which they could have been destroyed and created again, or in which the combination of then-existing particles led to new forms.

We should also note that the family tree shown is meant to be very schematic. Just as in the case of biological species, the development with time also led here to more and more different subspecies. But there is a difference: in the evolution of the early universe, the appearance of new species occurred through well-defined transitions. In the very beginning, primeval fermions interacted through

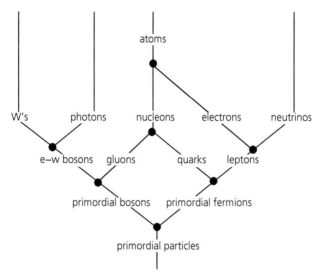

The family tree of the particle species

primeval bosons; there was only one primeval force. Why then did these primeval particles split into the multitude of the particle forms we know today? The only thing that changed was the temperature: the universe expanded and cooled off more and more. The decreasing temperature must have modified the structure of the world, just as it turns steam into water or water into ice. In Chapter 4, we shall consider such transitions in more detail.

At his point, we want to return briefly to a question that has occupied many physicists in the past few years. If we imagine that a certain transition turned the primeval fermions into a number of different quarks and leptons, why and how do they differ? We have labels to identify the different forms, charges and more, but we have not shown why there are different forms. Why are there six kinds of quarks or three kinds of electrons? This question reminds us of a classical earlier puzzle: why are there so-and-so many kinds of atoms? Could it not be possible that the basic constituents—quarks, leptons, and bosons—are just different combinations of something smaller and more "basic"? That is the aim of the so-called string theory, which has been studied in different forms over the past 30 or 40 years. It assumes that all our different particle species are just different vibrational excitations of tiny primordial strings. Such a picture is a continuation of the familiar reductionist approach, which so far has been very successful. It seems indeed tempting to identify six different quark species and six different lepton species as twelve different vibration forms of a single elementary string. That explanatory aspect, together with the fact that string theory is based on a challenging new mathematical formalism, is the reason why so many theorists have devoted their efforts to this approach. Unfortunately, however, these efforts have not yet led to a final and truly satisfactory answer. Present versions require a spacetime structure of more than the familiar four dimensions. The most popular form is based on eleven dimensions, of which seven appear much "smaller," just as the Earth seen from an airplane appears flat, since the heights of trees and buildings is so much less than the visible spatial extension. It is not so obvious, however, that a reduction in the number of particle species at the cost of an increase in the number of spacetime dimensions really is a genuine simplification. Moreover, such schemes predict, as we mentioned in the context of supersymmetry, the existence of further particle species so far not observed. The observation of such supersymmetric partners in future accelerator experiments would of course provide an immense boost to string theory as well.

We now have to return to two features so far left dangling: what is the snow we need to give our massless primeval fermions some mass, and where does the whole process take place? Our picture of the world so far lacks something that today is a crucial part of our universe: the so-called empty space, the stage for the whole play. The transition from the false to the correct ground state had liberated so much energy that the result was an extremely hot and dense medium of primeval particles. Everywhere there was something—the dominant, almost empty interstellar space did not yet exist. Let us now see how it came into existence.

3

Empty Space

In reality there exist only atoms in empty space.

DEMOCRITUS, 400 BC

Empty space is only an idealization, for various reasons; it is never truly empty. We are here not referring to the virtual particles submerged below the surface of the vacuum and waiting for the energy necessary to let them come up. They do not yet exist, and thus leave space indeed empty. But the empty space of our universe does contain the remaining dark space energy causing it to continue its expansion; beyond this, it does not create any structures or forms. This dark energy is, moreover, of extremely low density, much too low for any kind of particle creation, and it is distributed at constant density over the entire space. The counterpart of its density would correspond to about four nucleon masses smeared over one cubic meter—that's as empty as it can get. In addition, there is the cosmic background radiation, which also permeates all of space, though at a yet much lower and steadily decreasing energy density, decreasing because of the expansion of the universe. And the resulting "empty space" finally still contains us: it contains matter, which, however, is very unevenly distributed. If we average over the entire visible universe, we end up with a density of about one nucleon per cubic meter, less than that of the dark energy. Moreover, since matter is concentrated largely in stars and other heavenly bodies, the interstellar space is indeed very empty, as far as matter is concerned. In any case, it is the most empty space we can find in our universe, the *physical vacuum*.

At the start, however, in the very early universe, there was no form of empty space; the density of particles was immense; there was something everywhere. Even a spatial volume of a present nucleon, such as a proton, contained an immense number of quarks and antiquarks. There was no physical vacuum—even empty space first had to be created. The possibility for such a creation is due to the nature of the strong nuclear force, which binds quarks to form nucleons. Electrons are "real" particles: they can be isolated; it is possible to keep one single electron in a cubic meter of vacuum. With quarks, that is not possible: they are fundamental *constituents* and as such behave like magnetic poles. There is a plus pole and a minus pole, but these cannot be separated.

Quarks behave in a similar way. There are nucleons, as we have seen, in which three quarks are coupled together tightly. In addition, there are quark-antiquark pairs—in today's nomenclature, they are called *mesons*, strongly interacting particles that can be produced in high-energy proton-proton collisions. The symmetry determining the coupling of quarks to form nucleons or mesons is somewhat more complex than the plus-minus form of electromagnetism. The charge responsible for strong interactions can take on *three* rather than two different values, which are usually called *color*, such as red, blue, and green. Nucleons and mesons then arise through a superposition of such colors.

The fundamental theory of strong nuclear interactions, *quantum chromodynamics*, insists that only colorless objects can have an independent existence. Such color-neutral objects can be created either by combining a red quark with an anti-red antiquark (and similarly for blue and green) or by superimposing the three basic colors, which again leads to a colorless entity. An even simpler form of colorless state is a very dense plasma containing equal numbers of quarks of the three colors, as well as of quarks and antiquarks, all evenly distributed throughout. And that, we believe, was indeed the state of the very early universe: quarks and antiquarks everywhere, nowhere a spot of empty space.

The smallest possible color-neutral state then is either a quark and an antiquark, or three quarks, combined in a small spatial region so as to make that region colorless. In other words, quarks can exist in our present world, in empty space, only as quark-antiquark pairs of the same color/anticolor (*mesons*), or as three quark states whose colors are neutralized by superposition (*nucleons*). These are the smallest strongly interacting entities allowed to have an independent existence. They are collectively referred to as *hadrons*, from the Greek *hadros* (thick), to distinguish them from the very much lighter electrons, belonging to the set called *leptons*, from *leptos* (light). The interaction between quarks and between quarks and antiquarks is mediated by the force particles of the strong interaction, called *gluons*, since they "glue" quark constituents together to form hadrons. These gluons also have to be colored, so that they can bind different colors to form nucleons. Thus a red-green gluon is needed to couple a red quark to a green one, and so on, resulting finally in colorless hadrons.

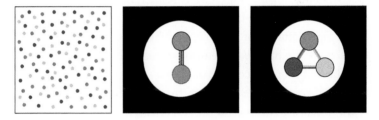

Quark states of existence: quark plasma, meson, nucleon

Quarks can only exist, as we have noted, if there are other quarks or antiquarks in the immediate vicinity, so that a local colorless combination is possible. Hadrons generally have universal size of about one femtometer (1 fm = 10^{-15} m); from this, we can conclude that this is the largest possible separation between quark companions. As Lucretius said, no force can separate them further. That also tells us what the lowest possible density of the quark plasma is: we have to have at least one quark per cubic femtometer. What happens if the density falls below that value? Then a new state of existence is created: our present world.

The early universe found itself in a state of continuous expansion and continuous cooling. The number of quarks could not increase, and so their density was necessarily decreasing. Sooner or later, a breaking point had to come: the quark density fell below that critical value of one per cubic femtometer; an unbound quark existence was no longer allowed. The overall situation now became a kind of cosmic "musical chairs": at the critical point, each quark had to grab an antiquark, or two other quarks, in order to survive as part of a colorless entity. And the hadrons thus created were now truly *free*— between two such hadrons there could be any amount of empty space; so this transition was indeed

The birth of the vacuum.

This birth also defined for the first time in our universe a measurable scale: the critical density at which a quark plasma becomes a gas of independent hadrons. This scale is defined by the limit of one femtometer (10^{-15} m), the largest possible separation between quarks. A larger separation is not possible; now hadrons appear, and between them there is what we call empty space. Hadron formation is thus also the first occurrence of *structure formation* in the universe: there are regions of empty space, and in it, from time to time, a bound state of three quarks. This type of process later was repeated a number of times: when nucleons coupled to form nuclei, when nuclei coupled with electrons to form atoms, when clusters of atoms led to the formation of stars, and when these eventually formed galaxies. In each case, structures appeared in empty space as a stage. How this can happen, we will address in detail in Chapter 6.

But the transition from quarks to hadrons has further consequences. At high density, the medium consisted of quarks and antiquarks, with gluons as force particles. The quarks had by then acquired a very small but still finite mass through interaction with the primeval snow given by some remaining underlying field (the dark energy or, as we shall see, more likely the Higgs field) permeating the entire universe—that was the mentioned "first" mass generation. The gluons remain massless and form for the quarks a much stronger background field, but one that exists only within the interaction range of the quarks. The same process of dynamical mass generation is now repeated—when the quarks move slowly enough through the gluon field, the gluons cling to them and give them a new, much larger mass. The gluons thus have to fulfill two functions: they give the quarks a rather large effective mass, and they mediate the interaction between these quarks, eventually binding them to create hadrons.

The quark mass created by the primeval space field was just a few MeV; the new mass, due to the cloud of the gluon field, is much larger, around 300 MeV, about a third of the proton mass. It is these, now much more massive quarks that combine to form hadrons. The nucleon mass thus lies around 900 MeV; the typical meson mass is around 600 MeV. In other words, the quark mass determined by the gluons as well the nucleon mass formed in turn by the heavy quarks are again emergent quantities, created by interaction; they are not inherent parameters of the system. And it is these masses we have in mind when we talk about masses and weights. The 80 kilos of a human due not arise from any primeval quark masses; they are due to the effective quark masses formed by the gluon field and by the nucleon masses resulting from them.

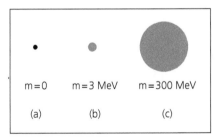

Evolution of the quark mass: (a) Primordial (b) Primeval field (c) Gluon field

In principle, the decreasing energy density of the medium could thus lead to two steps of evolution: first, the polarization of the gluon field results in new, larger quark masses; then, these massive quarks couple to form hadrons. All present calculations indicate, however, that these two phenomena occurred simultaneously in the early universe—it is not really well understood why that was the case. But as a result, we have a hadronization transition in which the quarks get their larger mass and are simultaneously bound to form hadrons. The plasma of colored quarks thus became a hadron gas: the *hadron era* began. In contrast to the colored quarks, which could never be further apart than one femtometer, there could now be arbitrarily much empty space between hadrons: the physical vacuum.

The birth of the vacuum was undoubtedly one of the four crucial events in the formation of our universe.

- The first was the fluctuation of the primordial world, leading to the exploding bubble subsequently becoming our universe.
- The second was the transition from the false to the true ground state, which liberated the energy needed to form the contents of our universe.
- The third was the breaking of the matter-antimatter symmetry, turning a world of equally many matter and antimatter particles into one in which matter dominates.
- And the fourth was the binding of massive colored quarks into color-neutral hadrons, making possible the existence of empty space.

The last step, the creation of effective gluonic masses for the quarks and their simultaneous binding to colorless states through

Hadrosynthesis,

led to what today is the major part of our world: space without any kind of matter. That is why we identified it earlier as the birth of the vacuum. But we have to be careful what we say—because space is continuously expanding, even at an increasing rate, it is clear that "empty" does not mean "nothing." There is, as noted several times, the dark energy inherited from the primordial world, out of which ours was formed. But this mysterious medium is in a sense a property of empty space—in any case, it is not something we would consider as matter.

What then was the form of the matter in our universe, just before and just after hadronization? The strongly interacting constituents, quarks and gluons, were dominant at first, making up some 80% of the total energy. It therefore makes sense to call the time before hadronization the *quark era*.

In the hadronization transition, the gluon field contribution turned into the effective quark masses and into the binding of the hadrons; we now had a mixture of hadron gas and lepton gas in empty space. At first, the hadron gas dominated, but hadrons and antihadrons could annihilate each other, leading to electromagnetic radiation. As long as the gas was hot enough, the inverse process brought new hadrons back into the system. With decreasing temperature, the process became more and more one-way, hadrons to radiation.

The hadron era ended as soon as the temperature was no longer sufficient to create hadron-antihadron pairs. Somewhat later, the leptons suffered the same fate, as we shall see shortly. And these annihilation reactions indicate a great potential danger: if the universe had contained an identical number of hadrons and antihadrons, in particular as many nucleons as antinucleons, the annihilation would have been complete, and today's universe would have contained no atoms, no matter in our present form. Our terrestrial world would never have appeared.

The tiny excess of quarks remaining at the end of the GUT era—one quark more among thirty million quarks and antiquarks—in the long run formed the whole basis for our world. The thirty million quark-antiquark pairs first formed hadrons and antihadrons, such as protons and antiprotons. These then annihilated each other and turned into radiation. But that one excess quark had no one to annihilate, so it had to find two further survivors to form a color-neutral nucleon, which now could irrevocably exist in empty space. The nucleons formed in this way were a tiny minority in space, which contained mainly electrons, positrons, neutrinos, and photons. The *hadron era* had ended; the *lepton era* started.

As we have already mentioned, the leptons shared the same fate only a little later on: electron-positron pairs annihilated each other, and only the minor excess of negative electrons remained from the end of the GUT transition, when it had compensated the equally small excess of positive quarks and thus kept the overall charge zero. The space of the universe now contained mainly radiation, photons and neutrinos, besides the few surviving nucleons and electrons. At this time, there were about a billion photons for each nucleon, so one could now rightly speak of the *radiation era*. And this ratio has remained roughly the same until now.

But in spite of its numerical majority, radiation no longer forms the main part of the universe. Space continued to expand, which increased the wavelength of the photons and thus lowered their energy, while the masses of the nucleons and electrons remained constant. The energy density of the photon gas thus decreased more and more, falling after some 50,000 years below that of the nucleons. The radiation era was now over; the domination of matter began. Today the energy density of nucleonic matter is more than a thousand times greater than that of the photon gas. In the following picture, we summarize the epochs of the universe mentioned; the size of the boxes does not, however, indicate the time extensions involved. The quark era was over after some 10^{-5} seconds, the lepton era after a further 10 seconds; the radiation era, as mentioned, persisted for about 50,000 years. And the matter era will be addressed in more detail shortly, since its fate involves us almost directly.

Big Bang	GUT	Quark	Hadron	Lepton	Radiation	Matter

Eras of the universe after the Big Bang

Let us return once more to the beginning of the radiation era. The content of the universe now consisted largely of photons and neutrinos, and in addition a few electrons and nucleons. The latter were protons and neutrons, with as many protons as electrons, to keep the overall charge zero. The temperature was now around 10^9 degrees

Kelvin; at this value, collisions with photons could still prevent the binding of protons and neutrons to form nuclei—any bound states were immediately destroyed again. Such break-up processes came to an end when the universe was more than ten seconds old. The energy of the photons was now no longer sufficient for break up, and nucleon binding became possible; in other words,

Nucleosynthesis

began, the binding of nucleons to form nuclei. At first, a proton and a neutron combined to form deuterium, then two deuterium nuclei formed a helium nucleus, and so on. Such fusion processes always involve two opposing effects: two nucleons attract each other through the strong nuclear force, but only at very short distances, while two protons repel each other electromagnetically, since they have the same charge. So to form a nucleus, the kinetic energy of the protons must be large enough to overcome the repulsion, and the overall density must be high enough to bring the nucleons close enough together. In other words, nuclear fusion can only take place in sufficiently hot and dense media. For this reason, all attempts to build nuclear fusion reactors have so far not succeeded—it has not been possible to create the necessary conditions in the laboratory. In the hydrogen bomb, nuclear fusion does become possible, because a prior atomic bomb explosion as a fuse provides the needed heat and density. In the early universe, the necessary conditions prevailed long enough to form at least helium. Before any abundant formation of much heavier nuclei, the world had already expanded and cooled off too much, so that after some minutes, the nucleons could no longer reach each other.

An essential feature of nuclear fusion, essential in particular for us here on Earth, is that the process liberates energy in the form of photons—electromagnetic radiation, sunshine. When two protons and two neutrons combine to form a helium nucleus, the mass of the nucleus is less than that of the sum of its constituents. The formation of a nucleus is therefore energetically more favorable than the persistence of the individual nucleons. That defines the temporal beginning of *nucleosynthesis*: it began, as mentioned, when the energy of the photons was no longer sufficient to break up the newly formed nucleus. Nuclei could now survive. And it ended when temperature and density had decreased enough: even when two protons now met, their electric repulsion would prevent nuclear binding.

Thus, there was only a small time window for primeval nucleosynthesis, nuclear fusion in the early universe; after a few minutes, the conditions were no longer suitable; and until then, essentially only deuterium and helium were produced. Because the requirements for nuclear fusion are known today, we can estimate what fraction of the nucleons were at that time bound to form nuclei: we expect about 25% helium, the rest remains as free protons, that is, hydrogen nuclei, and a very small amount of deuterium. It is found that those are indeed the relative abundances of nuclei observed in interstellar space; this provides one of the three pillars for the Big Bang theory.

The photons, which now made up a major part of the constituents of the young universe, thus arose from three sources. The first appeared when quarks and leptons went on separate ways, when the electromagnetic force split off from strong and weak nuclear forces. At this time, the bosons of the weak interaction acquired their huge masses, leaving the massless photons on their own. A further fraction was created in the annihilation of particles and antiparticles, hadrons as well as leptons—in both cases, the result was electromagnetic radiation. And a final fraction was due to the photons emitted in nucleosynthesis, the binding of nucleons to form nuclei.

These latter are for us of particular interest. The process of their creation, which first occurred as primeval nucleosynthesis in the chronology of the early universe, forms today the basis of our existence: nuclear fusion taking place in the interior of the sun sends photons down to our Earth and thus provides the light and the heat needed for our lives. What first happened in the early universe is today recurring in all the stars, including our sun: two protons and two neutrons combine to form a helium nucleus, whose mass is 6% less than the sum of the nucleon masses. This mass difference is converted into radiation, which becomes our sun light.

We were thus really lucky a couple of times. At the end of the GUT era, the specific fluctuation of our universe produced a few more quarks than antiquarks, together with a few more electrons than positrons. Otherwise, no matter could have survived in the further evolution—it would all have been converted into radiation. The next time, the expansion of our universe was just slow enough to provide the time window in which nucleons could bind to nuclei. A part of the nucleons were now united into nuclei, deuterium (2), helium (4), and very rarely some larger ones like lithium (7) and beryllium (9); some 75% remained as free protons, the future hydrogen nuclei. The mass anisotropy created by the heavier nuclei and their subsequent clustering later on allowed gravitation to form stars.

When nucleosynthesis ended, all the photons were still in continuous interaction with the charged constituents of the remaining medium, the electrons, protons, and nuclei. The universe was a hot, electromagnetic plasma. But in contrast to the quark plasma of still earlier times, there now existed empty space; there were regions containing "nothing." Nevertheless, a photon could not travel very far before it was scattered or absorbed by some charge. The electromagnetic plasma was completely opaque to light; our view back is stopped by this plasma just as our view into the sky is stopped by a layer of clouds.

The turning point came a little later, with a cosmological "little while": after about 380,000 years, the continuing expansion of the universe had lowered the temperature sufficiently to prevent photons from breaking up the atoms appearing through the binding of electrons with protons or with nuclei. The photon energy had decreased as their wavelengths increased, and they were no longer powerful enough to dissociate the newly formed constituents, the

Atoms.

The plasma of charged electrons and nuclei now turned into a gas of electrically neutral atoms. So as far as matter was concerned, there now were really only atoms, as Democritus has insisted. But besides such matter, there was lots of radiation, the remaining light of the Big Bang, to which we will return in a separate chapter. But first, we have to look in a little more detail at the different kinds of transitions.

4

Transitions

José Sobral de Almada Negreiros
Crossing the River Lima, Tapestry
Pousada de Santa Luzia, Viana do Castelo, Portugal

Commanded by Decius Junius Brutus, the Roman legions arrived in the year 135 AC on the left bank of the river Lima. Because of the beauty of the region, they thought to be on the banks of the legendary river Lethe, which destroys the memory of whoever crosses it. The soldiers therefore refused to go on. Carrying the standard of the Roman Eagles in his hand, the commander rode across the river to the other bank and from there called each soldier by his name, thus proving that this river was not the river of oblivion.

Transitions occur at the borders between different states of existence. You can change the temperature of water over quite a range, and nothing specific happens: water stays water. But then, suddenly, between −0.1 and + 0.1 degrees centigrade, the homogeneous, isotropic liquid becomes a solid, a crystal of ice with well-defined crystal axes and with completely different properties than those of water. Something similar happens between 99.5 and 100.5 degrees: now the smoothly flowing liquid becomes steam, in which the individual molecules have little or no communication, and which again has totally different properties than water.

In such transitions, the memory of the previous state is indeed completely eradicated. The water arising from melting ice has no information about its previous state, and no measurement can tell if the steam we observe was half an hour ago still water. The temporal evolution of our universe must have followed a similar pattern, with transitions between the different eras. We remember the problem of the vanishing antimatter: shortly after the Big Bang, matter and antimatter were still present in equal amounts, and then suddenly the antimatter was essentially gone. What kind of transition made that possible?

Before we address the transitions that took place in the early universe and led us to its present state, we want to consider in a little more detail the subject of transitions as such, as they are studied and described in physics. There are quite different forms of transition from one state of matter to another, and their description is a very interesting and rather novel area of physics. It's only fifty years ago that the American theorist Kenneth Wilson was awarded the Nobel Prize in physics for developing the foundations of the field, and a considerable part of its further progress became possible in the last decades through pioneering work carried out on huge supercomputers, that is, at large-scale high performance computing facilities. Let us look at some familiar concrete examples of transitions.

If with decreasing temperature precipitation turns from rain into snow, there are no sharp borders. Eventually a few flakes of snow begin to appear in what started as pure rain, and with time their fraction increases in comparison to that of rain, until finally it is just snowing. The constituents of the medium were transformed from rain drops to snowflakes. The transformation occurs with decreasing temperature, but continuously, over a whole range of values—in spite of the fact that the meteorological experts in Scandinavia are quite capable of distinguishing "snow-blended rain" from "rain-blended snow." But in this case we never have an abrupt confrontation of two different states, and so we prefer call transitions of this kind *transformations*: rain is continuously transformed into snow, and vice versa.

The evaporation of water, on the other hand, is something quite different. Up to 100 degrees the system is in a liquid state—we ignore the few escaping steam bubbles. If we now continue to add heat (and also ignore the quite unusual case of superheating), the temperature does not increase right away, but instead more and more of the liquid is turned into steam, until all the water has evaporated. The amount of heat required for this process, the "evaporation heat," is a well-defined quantity, and only after it has been delivered to the system does the temperature increase again. Until then we have a mixed state of water and steam; adding heat changes their relative fractions, brings the system at constant temperature from "only liquid" to "part liquid, part steam" to "only steam." The situation is quite similar for melting: the temperature remains constant at zero degrees, until the sufficient amount of "melting heat" has been supplied to turn all the ice into water.

A third transition form appears when metals become magnetic, and this form has had a decisive influence on our understanding of transition phenomena. Iron consists of atoms having their own spin; they seem to have an axis around which they rotate. If a material is magnetic, if it forms a magnet, it means that on average the spins of all the atoms point in a certain direction. At sufficiently high temperature, metals are not magnetic—the axes of its atoms are randomly oriented, so that an average over the directions of all atoms leads to zero—there is no preferred direction. For each atom, the spins of its neighbors point in random directions compared to its own spin orientation. If we now lower the temperature, clusters of atoms start to form, first small ones and then ever-larger ones, three-dimensional "islands," in which all spins point in the same direction. The overall spin directions of the different islands are not coordinated; they are again random. But if we now lower the temperature still further, something almost miraculous happens. At a certain point, at the *Curie temperature*, an island appears whose dimensions are as large as the whole system and in which all spins point in the same direction. There still exist other, smaller islands with randomly oriented spins, but with decreasing temperature they all join their big brother in their spin orientation. The final state is reached at zero temperature: now *all* spins point in the same direction. The arbitrary piece of iron has become a perfect magnet.

The process of magnetization

This transition occurs, as mentioned, at a certain well-defined temperature, bearing the name of the French physicist Pierre Curie. If we measure the average spin orientation of the atoms below this temperature, it is no longer zero; the overall spin points in some arbitrary but well-defined direction. There is a polar axis, from south to north, for the piece of iron as a whole. Pierre Curie started a famous, albeit tragic Nobel dynasty. He and his wife Marie Skłodowska were awarded the Nobel Prize in physics in the year 1903; it was the third ever given. Marie later on also received the Nobel Prize in chemistry in 1911; Pierre was no longer alive then, having died in a traffic accident in 1906, caused by a horse-drawn carriage. Their daughter Irene married one Frederic Joliot, and this couple was awarded the Nobel Prize in physics in 1935. Apart from Pierre, all members of the Curie family became victims of leukemia, caused by their work with the radioactive substances they had discovered in their pioneering research.

Marie, Irene and Pierre Curie

As we saw before this little detour, there are essentially three kinds of transition. The first, a transformation such as rain turning into snow, appears under numerous conditions and was also quite essential in the evolution of the universe. Nevertheless, it has not attracted the same interest in physics as the other two, because it happens continuously, as the relevant parameters—temperature, pressure, humidity, and more—pass through a range of values. Nothing happens abruptly.

Things look very different in the case of magnetization. This can best be illustrated by using a simplified model. Consider a regular grid, a lattice on whose intersection points there are spins of unit length pointing either up ($\uparrow$) or down ($\downarrow$): $s = +1$ or $s = -1$. Each spin is connected only to its nearest neighbor, and if we flip all spins simultaneously, the interaction remains unchanged. This model was studied by the German physicist Ernst Ising in his 1924 doctoral thesis, and it remains forever the Ising model, even though his thesis advisor Wilhelm Lenz had the idea and the actual solution was provided only in 1944 by the Norwegian-American physicist Lars Onsager. In any case, the Ising model today still constitutes one of the most important models in physics, one that has decisively formed fundamental concepts in wide areas of research.

At high temperatures, equally many spins point up and down, and both orientations are randomly distributed. With decreasing temperatures, the "islands" of parallel spins mentioned earlier start to form; there are some in which all spins point up, some in which all point down. The average over all spins remains zero. At the Curie point, however, at the temperature T_C, the system must decide spontaneously what to do: from now on, either up or down wins; both have the same chance. From now on, for all temperatures below the Curie point, the average over all spins is no longer zero; the majority points either up or down. In the following picture, we have illustrated the situation in the two-dimensional case.

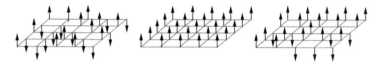

The Ising model: (a) above T_c, (b) just below T_c, and (c) at $T = 0$

This process is referred to as *spontaneous magnetization*, and it is clearly an *emergent phenomenon*: the individual spin is unimportant, and also its interaction with its neighbors alone cannot determine what happens. The effect sets in as a concerted action of all spins, and there is no specific trigger, no general giving a command. The observation of such a phenomenon is not quite as new as it may seem at first. In the biblical proverbs of Salomon (Proverbs 30:27) it is said

The locusts have no king, yet all of them go forth in ranks.

This is perhaps the first mention of something known today as swarm intelligence; it has become a new field of research in biology, and it plays a decisive role in the behavior of bird and fish swarms. In the inanimate world, it is in fact a basic phenomenon, as we just saw.

Another impressive example of the sudden change of behavior is considered under the heading of *percolation*. The simplest form is employed in the Asian game of Go. One randomly places stones on the cross points of a checker board. These first form small and then ever-larger islands. But suddenly, when some x-th stone is placed, the islands extend from one side to the other, and instead of randomly placed islands in a lake we now have randomly placed lakes in a connected land. Before the last stone was placed, there was no connection between the shores: this last stone suddenly created it. Such a transition is referred to as percolation—it is responsible for the fact that in a coffee percolator, the coffee all of a sudden begins to flow as we add more and more water. In the same way, the transition from the dominance of water to that of land on our checker board takes place very abruptly.

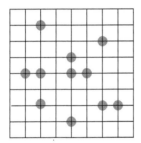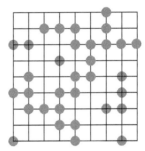

Percolation

In a two-dimensional world, the state is always "either or": either a sea with islands, or a land with lakes. Once the percolation forming land has taken place, it is no longer possible to reach all points on shore by ship. That is the case when less than half of the surface is covered by sea. The surface of our Earth, however, is about 70% made up of oceans, and so one can in fact get by ship from any shore to any other, as was shown five hundred years ago by the Portuguese seafarer Fernando Magellan.

For three space dimensions, that changes: the "either or" now becomes an "as well as." The holes in a Swiss cheese are bubbles, but we can imagine some worm eating its way from one side to the other, so that now there are "air channels" through the cheese, even though there still remains a cheese structure as well. A fence divides an area into two disjoint parts, but a tunnel does not do that to a mountain. We now quite naturally arrive at a mixed state, such as we had in the case of evaporating water. In the beginning there is the rock; then we drill tunnels through it, and continue drilling, until the rock collapses and there is no more rock structure, just stones and gravel.

But let's return to our spin problem. At the onset of magnetization, that state is not changing gradually with changing conditions. The average spin value above the Curie temperature is zero, and it is still zero at the Curie point; but from there on it is finite, no longer zero. This kind of behavior mathematicians call *singular* or *nonanalytic*. It always occurs when only a clear yes-no decision is possible. One cannot be a little dead, or a little pregnant, and so a system also cannot be a little magnetic; in percolation, the system cannot be a little interconnected. At a certain point, there is a crucial, qualitative change. In physics, such singular phenomena in collective systems are referred to as

Critical behavior.

Up to the critical point, measured observables were zero, and then no longer zero—or vice versa. In any case, the behavior of the system there changes qualitatively, fundamentally, abruptly, and not just a little or gradually.

At the critical point and in its immediate vicinity, for values of temperature and pressure close to the critical ones, the system can no longer be divided into smaller representative subsystems. Sufficiently far above or below the critical point, it is generally possible to study smaller parts of the whole in the expectation that the rest will behave the same way. In the critical region, that does not work anymore. In the case of spin systems, islands of all sizes appear, from two spins to clusters reaching from one side of the system to the other. For arbitrarily large systems, that means at the critical point the relevant scale, the correlation range, becomes infinite, it *diverges.* That is perhaps also the reason why such systems are so difficult to treat with our usual mathematics. Lars Onsager's solution of the Ising model was for the two-dimensional case; in two dimensions, the percolation problem is also solvable. In three or more dimensions, however, neither has so far been solved by analytical mathematics, in spite of many attempts. What we know about these cases comes from numerical studies on large-scale supercomputers.

Another very useful way of studying such transitions is based on symmetry. When iron is not yet magnetic, in its "paramagnetic" state at sufficiently high temperatures, the spins of the atoms are randomly oriented; the average overall spin value m is zero. In the Ising model, there are as many spins $+1$ as there are -1, randomly distributed. This result is not changed if we flip every spin orientation into its opposite, $+1$ to -1, South Pole to North Pole. That means the interaction connecting the spins with each other is not changed by such flipping; it cannot distinguish a state from its flipped form. Below the Curie point, however, at lower temperatures, there now is a preferred direction: there are more $+1$ spins than -1, or vice versa; m is no longer zero. The interaction is not changed at all but the state of the system has: flipping now changes the state into its opposite.

This situation is of great relevance for physics in general and in particular for the understanding of the evolution of our universe. If an interaction is unchanged by a certain operation, such as by flipping all spins, we call it *symmetric* or *invariant* under this operation. This does not mean, however, that the actual state of the system remains unaffected, as we have just seen. For this state, there are two possibilities: either it is also invariant, or it is in one of two asymmetric opposite states, either $+1$ or -1, and both of these are equally likely. This means that at the Curie point the symmetry is spontaneously broken, leading to $+1$ or -1, but because of the symmetry of the interaction, we cannot predict which of the two it will be. In some sense, this is like playing roulette: as long as the ball is rolling, it is with equal probability on a red or a black field. Once it stops, however, it has to decide, and in an honest game, the chances for either red or black are equal.

For a compact description of the situation, one defines the average m over all spins as the *order parameter*. Above the Curie point, there is complete disorder; as many spins point up as down; the measure

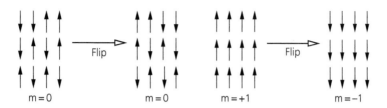

Flipping all spins above (left) or below (right) the Curie temperature

of order is zero: $m = 0$. If we now flip all spins, it remains zero. Below the Curie temperature, however, the order parameter is no longer zero; there is some kind of order. The average over all spins defines a specific direction, and when the temperature is zero, *all* the spins point in the same direction. Since the interaction does not depend on the spin direction, that can be up as well as down, but the system has to make a choice. If we now flip all the spins, +1 becomes −1 and vice versa: the actual state of the system is not invariant under flipping, even though the interaction is. One calls the process occurring at the Curie point *spontaneous symmetry breaking*. It occurs even though no one or nothing intervenes. There is no change whatsoever in the interaction, but nevertheless the system suddenly falls into an asymmetric state. This self-induced character is what distinguishes the phenomenon from *explicit symmetry breaking*, which occurs, for example, if we break off one point of a six-cornered star.

We have so far not paid much attention to the form of transition encountered when water evaporates or ice melts. But in the discussion of percolation, we saw how such a process can occur: at the transition point, both states persist, water and steam, or water and ice. There we can, as an example, consider water as ordered and steam as disordered, using the density of the medium as order parameter. In this case, the system passes gradually, *at fixed temperature*, from an ordered to a disordered state. While the order parameter for magnetization decreases with temperature continuously to zero, it drops in the case of evaporation from a finite value to zero at a fixed temperature. Below the evaporation temperature, it is different from zero, and it remains finite as further heat is added. The input of energy increases the fraction of steam relative to that of water and thus reduces the value of the order parameter, which finally vanishes when everything has become steam. In contrast, for magnetization the order parameter vanishes smoothly as we approach the critical point from below, and so one calls such transitions *continuous*, whereas evaporation and melting are *discontinuous* transitions. In the following picture, we illustrate the behavior of the order parameter in the two cases. Here $m(T)$ is average value of the spin as function of the temperature, while $d(T)$ is the density of the medium minus the steam density.

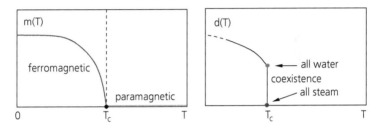

Order parameter for the continuous magnetization transition (left) and for the discontinuous evaporation transition (right)

With that, we have a summary of the different forms of transition from one state of matter to another. It can happen gradually—transformations without any specific indicative parameter values; we considered the change from rain to snow as an example. Transitions can occur at a fixed temperature, at which two different states coexist; further energy input then shifts an ever-larger fraction from the old state to the new; here we looked at the evaporation of water as a typical case. And, finally, the transition can occur unambiguously, at a precise parameter value, as seen at the onset of magnetization: at the Curie temperature, the whole piece of iron, paramagnetic up to that point, from now on is a ferromagnet: the metal suddenly becomes magnetic.

What does all that have to do with the evolution of the universe? We have seen that the early universe was hot and dense, and that it subsequently cooled off and became more dilute. That originally it contained equally many particles and antiparticles, but at the end only particles. That originally all particles were massless, but some suddenly acquired masses. That quarks combined to form nucleons, nucleons and electrons to form atoms. So the chronology of the universe contains quite a few different transitions, and we now want to look at some of them in more detail.

First of all, we recall that at a transition, different physical processes can occur. The symmetry form of the state can change, from $m = 0$ (symmetric) to $m \neq 0$ (asymmetric). The building blocks of matter can change, from massless to massive. And they can combine to new, more complex building blocks, like quarks to nucleons.

If we ignore the presently still very speculative possibility of one single primeval form of particle, then we start with two particle species, primeval fermions and primeval bosons. The first transition of importance for us, then, is the one at which the primeval fermions split up into quarks and leptons. With the decreasing temperature of the continuously expanding universe, a point was reached, at some 10^{27} degrees Kelvin and about 10^{-35} seconds after the Big Bang, at which each fermion had to choose one of the two possibilities: the "grand unification" was over. Until then, transformations from one form to the other were possible; from then on no longer. Because of such transformations, the number of quarks could deviate temporarily from that of antiquarks, and correspondingly that of leptons and antileptons. At this GUT transition, the basis for the present asymmetry between matter and antimatter must have been created: at this point, a symmetry must have been spontaneously broken, just as the symmetry under spin flipping in the Ising model mentioned earlier. Up to the transition point, the fermion species were indistinguishable, and their number was equal to the correspondingly indistinguishable antifermions.

From then on, however, there existed two distinguishable species, quarks and leptons, as well as their antiparticles, and now neither the quark-antiquark difference nor that of leptons and antileptons had to be zero: small, but not zero. These numbers therefore constitute order parameters, and their finite values mean that the GUT symmetry given by the grand unified interaction was from now on spontaneously broken by the actual state of the universe.

Because later on the quarks became the basis for nucleons, the asymmetry was an absolute prerequisite for our present world. As we shall shortly see, in annihilation reactions between matter and antimatter, only a possible excess of one or the other kind will survive, and so without such an asymmetry, the universe today would only contain radiation and no matter. That is, so to speak, the other side of the coin. If energy is deposited in empty space, it gives a pair, fish and antifish, the possibility to surface to reality. But if subsequently fish and antifish should meet again, they can annihilate each other and thus turn into electromagnetic radiation. For this reason, a universe containing exactly the same amount of matter and antimatter would in the long run dissolve into radiation. There had to be a point at which the matter/antimatter symmetry was broken.

We mentioned that the charge distinguishing between matter and antimatter, between quark and antiquark or between nucleon and antinucleon, is denoted as baryon number. The baryon number of the nucleon is $+1$, that of the antinucleon -1. The transition at the end of the GUT era thus becomes

Baryogenesis,

the creation of what in the long run became a matter-dominated universe. The trigger for this abrupt transition is found in the nature of the force particles. Up to this point, all fermions were treated equally, and hence also all force particles were equivalent and massless. There were those mediating between quarks, those mediating between leptons, and those changing a quark into a lepton. All these force particles were massless, and all the interactions they initiated were equally strong.

The end of this egalitarian world arose when the force particles changing a quark into a lepton or vice versa (they are usually referred to a X bosons) suddenly acquired an extremely large mass.

Such a process is possible in the GUT forms studied so far; it signals the spontaneous breaking of a symmetry of the interaction and thus corresponds to a critical transition, similar to the onset of spontaneous magnetization. How such a mass-creating transition can occur was discussed in Chapter 2: it is yet another case of the snowball syndrome, in which the rolling stone gathers moss after all, provided it does not roll too fast. But what is the moss in this case? That is a crucial question, so far not really satisfactorily answered. We shall therefore put it aside for now and return to it when we discuss the next, so-called Higgs transition. In any case, the huge mass increase means that the X bosons are effectively removed from the scene. Two fermions could no longer interact by their exchange, and hence the transformation of a quark into a lepton or vice versa was effectively excluded from now on. Neither quarks nor leptons have thus changed in any way: both kinds are still massless. But by suppressing the force particles that could change one species into the other, the unified world of the species was broken. From now on, there were quarks and there were leptons, and it was impossible to change one into the other.

We noted that quark-lepton transitions are *effectively* excluded, because the large mass of the X bosons suppresses such interactions. Although they are extremely unlikely, they nevertheless remain in principle possible: the X-boson mass became huge, but not infinite. So quark-lepton transitions are not strictly forbidden: there is a minute probability for a quark to turn into a lepton and then subsequently find an antilepton to annihilate and turn into radiation. That way our world could still eventually disappear—its existence is based on the stability of quarks and in turn on that of protons, and the possibility of quark-lepton transitions imply proton decay.

In the framework of GUT, the expected lifetime of protons has been calculated to be some 10^{32} years. So one really has to wait a long time until a given proton decays, particularly if we keep in mind that our universe is only 10^{10} years old. But it's not that simple. A human is made up of about 10^{29} protons, so that a tank of water with a volume of 1000 humans should produce one proton decay per year. Such experiments have been carried out and continue to be in progress, with big tanks of water in old mines and similar places, to eliminate as far as possible any effects of external radiation. A well-known setup exists in the Japanese community Kamioka, where deep underground a tank of 3000 tons of pure water is surrounded by 1000 photo detectors. If a proton should decay in the tank, the resulting electron would emit an observable light signal. So far, nothing has been seen.

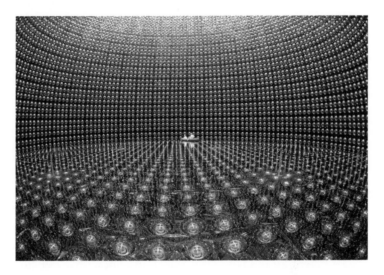

The interior of the Kamiokande setup. Photo: Kamioka Observatory, ICRR, U. of Tokyo, Japan

As far as matter particles are concerned, at the end of the GUT era, the universe contained on the one hand a dense medium of quarks and antiquarks, and on the other an equally dense medium of electrons, positrons, neutrinos, and antineutrinos. Interactions between the constituents of the two groups were still possible, but no more transitions from one to the other, with the caveat just mentioned. And in both groups, there was a tiny asymmetry, a little more matter than antimatter, and correspondingly a few more electrons than positrons, to keep the overall charge zero.

How does evolution now continue? First, we should emphasize once again that the present considerations concerning the unification of the particle species so far remain speculative. There are, as indicated, some 44 or more different models, which more or less contradict each other in some aspects; nothing is yet final. And testable predictions, such as those for proton decay, have not been confirmed by experiment. If one should find in the near future evidence for proton decay, that would of course really mean progress. But until then, the considerations mentioned indicate only how it might have been.

Things become more definite as we progress to somewhat later, but still very early times; we now enter a region in which terrestrial experiments can indeed contribute significantly to our understanding. So far, all matter particles were massless, and the force particles were as well, except for the X boson, which now appears to be unimportant. We are now confronted by the question of what remains of the false ground state after the transition to the true present one. There still is the dark space energy, which continues to expand the space of the universe. That question is even today not answered in a fully satisfactory way. As we have known for about two years, there is a further mysterious medium penetrating the entire universe— the *Higgs* field. It is in nature quite similar to the dark energy, and there are speculations relating the two. The Higgs field, however, has a different raison d'etre: it is the cosmic snow that can clump to the bare primeval particles and thus give them masses. Below a certain temperature, the Higgs field changes the nature of the force particles of the electroweak interaction; so far, there were four, all massless. In the Higgs transition, three of them suddenly acquire very large masses; the fourth remains massless and is from now on called a *photon*. For this reason, the Higgs transition is in a sense

The birth of light.

From now on, there is electromagnetic radiation on one hand and radioactive decay on the other. The world of leptons is thus split into two sectors: purely electromagnetic interactions, mediated by massless photons, and the so-called weak nuclear force, responsible for radioactive decays. It is precisely the new, large mass of the bosons mediating this interaction that makes it so weak and so short-ranged. Moreover, the Higgs field also clings to the matter particles and gives them small, albeit finite masses, in a similar way as seen for polarization in a plasma. The electrons and positrons, so far massless, now acquire their intrinsic mass, as do the quarks. It is claimed that the neutrinos suffered a similar fate—but so far, their mass has not been unambiguously determined. The interaction between quarks, however, is unaffected—the gluons remain massless.

So after the GUT transition, we have encountered another change of state, the electroweak or Higgs transition. At the GUT transition, quarks and leptons separated; at the Higgs transition, electrons and neutrinos go separate ways, and from now on, all matter particles have small but finite intrinsic masses. At the Higgs point, photons and W bosons also separated: the latter share the fate of the X bosons: they acquire such huge masses that from now on, they are out of the game. This transition occurred about 10^{-10} seconds after the Big Bang, when the temperature had dropped to some 10^{15} degrees Kelvin. We have to emphasize here that the masses created in the Higgs transition have very little to do with what we today refer to as masses in our everyday world. Our weight is determined by the atoms in our body; more precisely, by their nuclei. The masses of the electrons don't play any significant role, and the intrinsic masses of the quarks don't, either; we come back to that shortly. For the world of the leptons, the Higgs transition was the last: both electrons and neutrinos have made it unchanged into our present world. And the photons have as well, of course.

Before we turn to the fate of the quarks, let us come back to those three mysterious fields that permeate everything: the remaining dark energy causing the ongoing expansion of the universe, the GUT-field that gave rise to the large X-boson masses, and the Higgs field making the W bosons so massive. The latter should actually be called the Higgs-Brout-Englert-Guralnik-Hagen-Kibble field, since all these theorists more or less at the same time proposed and investigated such a snowball form of mass creation. We certainly cannot resolve the priority issue; in any case, the Nobel Prize 2013 was awarded to Englert and Higgs; Brout was no longer alive at that time. For simplicity, we shall join most others and remain with Higgs—all the more so because Peter Higgs is an extremely modest, very academic scientist, who never clamored for the fame today assigned to him. He had concluded that the existence of the field now named after him must also imply the existence of an associated heavy boson; moreover, this boson should be observable in high-energy collisions of elementary particles. By now, it seems clear that in 2012 this particle was in fact first observed at the European Center for Nuclear Research (CERN) in Geneva—and that led a year later to the Nobel Prize for Englert and Higgs.

François Englert and Peter Higgs, 2014
Photo: CERN, Geneva, Switzerland

The next in the line of transitions is one that already affects us quite directly. The quarks, through their interaction, give rise to what we today call the *strong nuclear force*—and that leads to the binding of quarks to form nucleons. This *hadronization transition* is also, as we have seen, the basis for the creation of the physical vacuum, empty space as such.

Before this transition took place, the universe was densely filled with quarks and antiquarks; no quark was ever more than a femtometer (10^{-15} m) away from another one. At the hadronization transition, three quarks then combined to make a nucleon, three antiquarks an antinucleon, and a quark-antiquark pair a meson. And these new entities, the hadrons, could exist alone, all on their own, in empty space, without other hadrons nearby. Once more we encounter here a transition similar to that in the Ising model: one state, the quark-gluon plasma, becomes a gas of hadrons. The plasma consisted of unbound color charges, and if we imagine a "color voltage," then this medium was a color conductor—there could be color currents. At the hadronization point, that ended; color conductivity now dropped to zero, so that we again have an order parameter that is finite in a certain range of temperatures and then vanishes in another range.

But, as we have already indicated, that was not all. In the plasma, the quarks were effectively massless: they had only the tiny intrinsic mass due to the Higgs field. At the hadronization point, gluon clouds formed around every quark, and these clouds gave to the quarks what we now, in our world, call mass. Each quark acquired about a third of a nucleon mass, some 10^{-27} kg. Three such quarks then formed a nucleon, and about 10^{29} such nucleons make up the weight of our body. That weight is thus made up almost totally of the gluon cloud surrounding otherwise very light quarks.

The effective quark mass,

created by the gluon cloud, again constitutes an order parameter: it is (almost) zero in the plasma and then suddenly becomes finite.

The hadronization transition, at which the universe for the first time began to look a little like it does today, occurred about 10^{-5} seconds after the Big Bang, at a temperature of some 2×10^{12} degrees Kelvin. The determination of the time depends a little on the specific model for the cosmic evolution, but the transition temperature is quite well known by now, from theoretical calculations in the context of quantum chromodynamics, as well as through nuclear collision experiments, in which a quark-gluon plasma is created in the laboratory for a few fleeting moments.

Our universe now consists of "empty" space—keeping in mind the caveats concerning "empty"—and of particles, mainly hadrons and antihadrons, electrons, positrons, neutrinos, and photons. These constituents form a hot gas in violent interaction, with continuous annihilation and creation. Before turning to the further fate of this gas, we first summarize in the next picture the transitions in the early universe and their effects on matter particles.

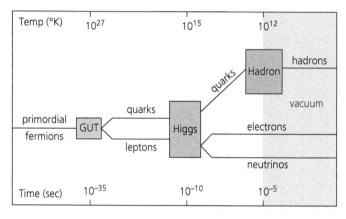

The transitions in the early universe and their effects on matter particles

But, as already mentioned, the different transitions also had their effects on some of the force particles and modified them, which then in turn changed the form of the interaction. We summarize these changes in the next figure. At the GUT transition, the primeval bosons gave rise to the gluons of the strong nuclear interaction and the leptonic vector bosons, which govern the electroweak interaction. At the following Higgs transition, the latter set split into photons and the so-called W bosons. Gluons and photons remain massless, while the clouds of the Higgs field create the large masses of the W bosons and thereby assure that the weak nuclear force is indeed weak and short-ranged. In the last of the transitions, hadronization, the gluons play a twofold role: on the one hand, they form the clouds that give the quarks and hence later the hadrons their observed masses; on the other, they bind these massive quarks to today's hadrons. The interaction between the hadrons is now taken care of by the mesons, whose masses determine the observed range of the strong nuclear force.

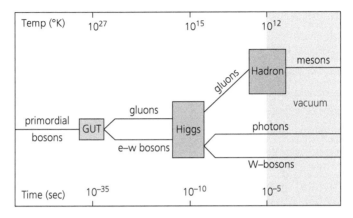

The transitions in the early universe and their effects on the force particles

The world created by the hadronization transition was quite short-lived. All matter particles were in continuous interaction; hadrons and antihadrons annihilated each other in collision and were recreated in other collisions; electrons and positrons shared this fate. But the expansion of the universe let the temperature decrease quickly, and with it the momenta of the constituents; as a result, the world soon reached a state in which the production of hadrons through collision was energetically no longer possible. This meant that in a very short time, almost all hadrons and antihadrons annihilated each other, producing photons and electron-positron pairs. The only survivors in this mass annihilation were the few excess nucleons existing from the end of the GUT transition. Up to the hadronization point, the universe was still almost symmetric, as far as matter and antimatter were concerned—but fortunately only almost. Now all antinucleons destroyed each other in collisions with nucleons, leaving only the few lucky excess nucleons. They were the entire future matter of our world; since then nothing more was added. Thus the nature of the universe was completely changed soon after the hadronization:

except for those few nucleons, the entire sector of world originally consisting of quarks was now fully converted into radiation and leptons. The quark era was over; the universe now consisted largely of leptons and photons: the lepton era had begun.

But its duration was soon threatened by a similar fate as that of the quark era. In the beginning, the world contained leptons of immense number, again *almost* as many electrons as positrons—the excess of electrons was just what was needed to keep the world as a whole electrically neutral, to balance the charges of the few remaining protons. This meant, as before, one electron among many millions of electron-positron pairs. Things remained that way as long as the temperature was high enough to compensate pair annihilation through pair creation. But the expansion continued, the temperature went on decreasing.

And this meant that eventually a temperature limit was reached, below which electron-positron pairs could only be annihilated—the radiation energy was too low for pair re-creation. In a short time, all the electrons and positrons that had determined the lepton era had left the scene; the only survivors were again those that made up the excess at the GUT transition. The excess of electrons and quarks left over from the earliest phases was thus to become the basis for the entire future world.

We have here ignored the neutrinos: for a while, they could still interact with the remaining nucleons and electrons, for example, converting a neutron into a proton and an electron: $\nu + n \rightarrow e^- + p$. But soon the temperature became too low for even that, and the neutrinos were thus the first particles released into complete freedom, decoupled from all remaining matter. They could now propagate unconstrained throughout the entire universe, and as such they became a kind of vanguard for the light of the Big Bang, which we consider in the next chapter.

At this time, some ten seconds after the Big Bang, and at a temperature of about 10^9 degrees Kelvin, the lepton era was over. Since electrons and positrons had largely annihilated each other, as hadrons and antihadrons had done before, the universe now contained almost only radiation, and number-wise, this would never change again. Per nucleon there were some 10^{10} photons, so one can now truly speak of a radiation era. The universe was still a hot plasma, in which protons, electrons, and photons continued to interact. The neutrons, because electrically neutral, could only participate indirectly: since they are more massive than protons, they decay after about 15 minutes into a proton, an electron, and an antineutrino. The relative numbers of protons and neutrons had first been determined at the end of the lepton era, when neutrinos were still able to convert one into the other. Because neutrons are slightly heavier than protons, they were at a disadvantage; their ratio converged initially to an average of about 15 neutrons to 85 protons. But the subsequent neutron decay would have led to complete extinction, had not a salvation mechanism entered the scene, as we shall see shortly.

Since the hadronization transition, we have seen the universe pass through a number of different states, from a hadron gas to a lepton gas and then to a radiation-dominated electromagnetic plasma. In the terminology we introduced at the beginning of this chapter, these were transformations rather than the sharp critical transitions encountered in the very early universe. Also, in the last of states mentioned, the electromagnetic plasma, once in a while a proton and electron would join to form a neutral entity, a first atom. Not for long, however: the energetic photons of the plasma would soon break it up again. But now something else set in: a proton and a neutron joined to form a first nucleus, deuterium. The binding is achieved by means of the strong nuclear force, and the mass of the deuterium is slightly less than the sum of the proton and neutron masses. The energy gained in this way is then emitted as radiation, through the reaction $n + p \rightarrow D + \gamma$.

As long as the energy of the photons in the plasma was sufficient to break up such a nucleus, the binding was not for long. But once the plasma temperature had fallen below 10^9 degrees, the photons were no longer strong enough to do that: the era of nucleosynthesis, fusion, had started. And the process continued: two deuterium nuclei joined to form a helium nucleus, again slightly less massive than the sum of its constituents. And such a nuclear fusion process could in principle continue, resulting in lithium, beryllium, and so on. Because the masses of the nuclei were always less than the sum of the constituent masses, the neutrons now were too light, so to speak, to still decay: the neutron mass in a nucleus was not enough to make a proton and an electron. This was the previously mentioned way out for the neutrons, the mechanism allowing them to survive.

But such nucleosynthesis also turned out to have only a finite time span to proceed, and again it was the expansion of the universe that was responsible. Nuclei consist of protons and neutrons, and the positively charged protons repel each other electrically. Fusion is therefore possible only if the momenta of the nuclei are large enough to overcome this repulsion. In other words, once the temperature of the plasma had decreased to a point at which the proton momenta were no longer sufficient to allow them to get close enough to each other for fusion to set in, nuclear binding was over. The process could therefore take place only in a small time window: the photons of the plasma had to be "soft" enough to prevent them from breaking up the nuclei, but the protons had to be "hard" enough to get near each other in collisions, to overcome the electric repulsion. In addition, the expansion reduced the density of the universe and thus also the chance for nucleons to meet.

Before we turn to the further fate of the photons, we want to look back once more and list the main forms of transition that we encountered in our time line of the early universe.

- The first transition occurred when the energy of the primeval bubble in the multiverse fell from its false ground state to its right normal one, comparable to the escape of a bubble of steam from superheated water.
- The second took place at the end of the GUT era, when the force particles, which until then could change quarks into leptons and vice versa, suddenly acquired huge masses, ruling out such transformations. Quark and leptons now were "different."
- The third happened during the quark era, when the force particles responsible for the weak nuclear force suddenly became

so massive that the force in turn became very weak and short-ranged. In this electroweak or Higgs transition, quarks and leptons acquired small but finite "intrinsic" masses.

All three transitions were triggered through mysterious primordial fields permeating the entire world. In the first, the difference in energy between the false and the true normal state of the *inflaton* field gave rise to the creation of matter. In the second, one expects that the energy difference between such false and true states of a field so far without a name gives rise to the large masses of the X bosons and thus to the splitting of primeval fermions into quarks and leptons. In the case of the third, the energy difference between the normal states of the Higgs field at high and low temperatures leads to the intrinsic masses of fermions and W bosons.

The Higgs field is the only one among these three fields that now appears to be quite well established: after the transition, there remains the heavy Higgs boson, which was finally seen in 2012, created in high-energy proton-proton collisions at CERN. The production of a corresponding inflation boson for the first transition seems forever out of reach for terrestrial experiments. Nevertheless, the continuing and even-increasing expansion of the universe caused by this field is experimentally confirmed. In the case of the second, the source of the X boson, we can only note that there is ongoing intensive search.

- The fourth transition in the time line is of a basically different nature: it has colored quarks joining to form colorless hadrons, which can exist individually and thus allow the appearance of the physical vacuum, space without matter.

In this case, the responsibility is not attributed to some mysterious medium permeating the entire universe. The crucial role is here played by the force particles of the strong interaction, the gluons: they give rise to the effective mass of the quarks through a gluonic snowball effect, and they bind these massive quarks to form hadrons. This transition is today being studied with increasing precision on both sides—before and after hadron formation—and it is investigated experimentally in high-energy nuclear collisions.

The four transitions we just considered were indeed critical phenomena, in the sense of critical behavior as we have defined it. In
each case, a symmetry effective up to that point was spontaneously
broken; the universe fell from a state of higher to a state of lower
symmetry. Our picture thus corresponds to an initial, primordial
universe that was maximally symmetric, and in the course of its
evolution this symmetry was reduced more and more through
spontaneous breaking. To have a pictorial illustration: we start with
a three-dimensional sphere turning into a two-dimensional circle,
which in turn is transformed into an octagonal star and finally into
a square. The sphere was invariant under all spatial rotations, the
circle only under rotations in a plane; the octagonal star remains
unchanged under eight rotations of 45 degrees each, and the square
only under four 90-degree rotations. Something similar happens
when water freezes to become ice: the interaction between the water
molecules remains invariant under all spatial rotations, but the ice
state only under a finite number (like four) of rotations through a
finite angle around a crystal axis. The symmetry of the molecular
state of water is spontaneously broken when the liquid freezes.

After the earlier phase transitions and the subsequent transformations, the world now consisted essentially of light nuclei, including
protons as hydrogen nuclei (the remaining neutrons had decayed),
and in addition electrons, photons, and neutrinos. The latter were
already liberated, because of their extremely weak interaction with
other constituents. What remained was thus a still rather hot plasma
of nuclei, electrons, and photons. Since both nuclei and electrons
carried electric charges, the photons were in continuous interaction
with both—they could not (yet) escape.

Nuclei and electrons were able to combine to electrically neutral entities only when the temperature had decreased still further:
now the first atoms entered the stage, and the universe had reached
a state in which it still is today. The time between the beginning of
nucleosynthesis and the formation of the first atoms was truly an
eternity as measured in the scales of the early universe: only some
380,000 years after nucleosynthesis were all nuclei and electrons combined into atoms, and the universe contained only electrically neutral

constituents. Because photons can only interact with electric charges, they now had no more counterparts; they were free and could spread out unconstrained. We can thus slightly extend the words of Democritus: "In reality, there exist only atoms, radiation, and empty space," including the neutrinos in the radiation. While the amounts of matter and radiation are fixed forever, the empty space is not static— thanks to the remaining dark energy, it continues to expand. That has a crucial effect on radiation, as we shall see in the next chapter.

First, however, we want to combine into one picture all these many steps taken by the universe since the Big Bang. It is divided into an early, extremely hot and dense part, before the appearance of the physical vacuum, and another part, still rather hot and dense, but already containing what we call empty space, the stage for all that came since then and still exists today.

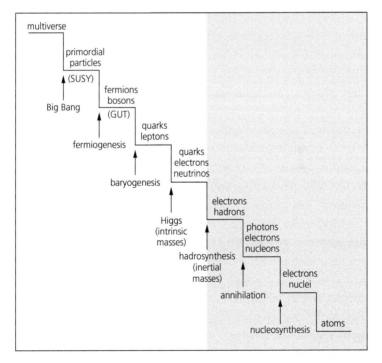

The stages of the universe from the Big Bang to the formation of atoms; in the blue region the physical vacuum already exists

In this picture, we only show the succession of steps as they occurred in the evolution of the early universe. But it is of course also of interest to learn when what happened, at what "time after the Big Bang," and what the state of the universe was at that time, "how hot" it was. The answers to these question obviously become less and less certain the further back we go. We try it anyway, and we begin with the time line.

As we have mentioned several times, the present atomic universe was born some 380,000 years after the Big Bang; that was when electrically charged nuclei and electrons combined to form neutral atoms. Before this time, there was the radiation era in which unbound nuclei and electrons together with photons and neutrinos were the essential constituents. This era had started some ten seconds after the Big Bang, because at that time most electrons and positrons had annihilated each other, turning into radiation, except for those electrons needed for today's world. The hadron era before this had ended about one second after the Big Bang, when nucleons and antinucleons suffered this fate, annihilating each other except for what is left today, the rest becoming radiation. A large part of today's photons was thus born in these two transitions, nucleon-antinucleon and electron-positron annihilation. The hadron era itself began in a genuine transition, 10^{-5} seconds after the Big Bang, when the density of the universe had dropped to a point at which the required quark density risked falling below that needed to assure each quark its partner. Quarks and antiquarks now were forced to combine to hadrons and/or antihadrons.

This transition, as mentioned, is the first that is quite well understood both theoretically and experimentally, through calculations based on quantum chromodynamics (the theory of strong nuclear interactions) and though experiments studying high-energy nucleus-nucleus collisions.

Before this, we have the Higgs transitions, which gave quarks and leptons their small intrinsic mass and the weak bosons their large ones. The determination of the mass of the Higgs boson in experiments at CERN has also here provided an empirical basis.

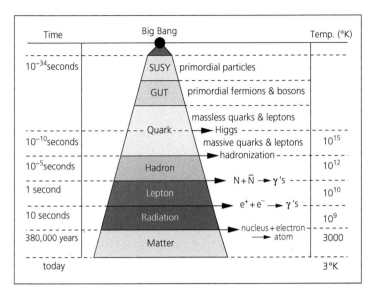

Time and temperature scales in the evolution of the early universe

The situation at still earlier times is quite uncertain, specula-tive, and dependent on the theoretical basis adopted. One tries to extrapolate the energy dependence of the couplings for strong and electroweak interactions to a point, at which they become equal: that would define the GUT transition, at which primeval fermi-ons separated into quarks and leptons. Present considerations lead to the values shown in the previous figures. Comparing the time of the GUT transition shown there with the end of the inflation as discussed in the first chapter seems to have the GUT transition occurring before the end of inflation. That cannot really be possible, because only at the end of inflation, after the transition to the true ground state, did the energy for the creation of particles become available. The numbers simply are not consistent. We can only note here that neither the 10^{-34} seconds for the end of inflation nor the 10^{-35} seconds for the GUT transition are known well enough to draw any conclusions. And if we believe supersymmetric considera-tions, there should even have been a yet earlier transition, in which the primordial constituents were split into fermions and bosons. Our understanding of such a transition, as that of the GUT transition, is today simply not adequate. The other scales, as estimated, are included in the following picture.

At this point we should emphasize once again that the picture we are presenting here is that corresponding to the "new" cosmology, based on the work of Alan Guth, Andrei Linde, Paul Steinhardt, and others. It has our universe appearing as just one of the many bubbles continuously created in the multiverse. The appearance of the bubble and the subsequent inflation are now physical phenomena. The "old" cosmology was based on general relativity and its temporal and spatial singularity, constituting a unique Big Bang outside all "normal" physics. Following this singular event came the Planck era, in which all forces, including gravity, were unified. In such a picture, the separation of gravity and inflation occurred still sooner after the Big Bang, after about seconds. On the one hand, this "old" view suffers from the absence of a quantum theory of gravitation, a quantum theory combining strong and electroweak forces with gravity. It is not evident whether such a theory would still lead to any singularity. On the other hand, it seems not clear today if gravity, as a possibly "emergent" force, has any place in such a unified theory. Conceptually, at least the "new" cosmology provides a rather satisfactory framework, albeit at the cost of introducing worlds outside our experimental reach.

5

The Light of the Big Bang

And God said, Let there be light, and there was light.

THE BIBLE, GENESIS I.3

The light of the Big Bang is today no longer visible to our eyes; nevertheless, it is present everywhere, uniformly distributed throughout our universe: it the cosmic background radiation already mentioned, discovered in 1964 by Arno Penzias and Robert Wilson. The wavelengths of visible light range from the ultraviolet (4×10^{-7} m) to the infrared (7×10^{-7} m). The 2.7 degree Kelvin cosmic radiation, in contrast, today has an average wavelength of about half a millimeter and thus falls into the microwave range. But when it was first liberated, some 380,000 years after the Big Bang, its wavelength was a thousand times less, in the yellow-orange range of our visible light. So at that time, the sky was not dark at night, but bright yellow! Shortly afterwards, the expansion of the universe moved the wavelength of that light into the invisible for us, but if some creatures were to exist with sensors for these wavelengths, then they would find that even today the night sky is not dark. In fact, we do have access to such sensors: in the absence of a signal from some station, our TV screen registers that well-known milky noise. A small part of it, about 1%, is due to the cosmic background radiation. So at frequencies between two stations, we receive the message from the Big Bang.

In the last chapter, we showed that about 380,000 years after the Big Bang, when electrons and nuclei had combined to form electrically neutral atoms, there was nothing left with which photons could interact in any way. From this point on, for them the world was transparent. So the cosmic background radiation we

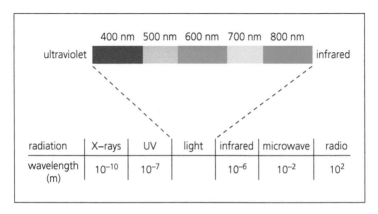

The spectrum of electromagnetic radiation (1 nm = 10^{-9}m)

observe today consists of the photons liberated at that time; they are unchanged since then, except that the subsequent expansion of the universe has increased their wavelengths correspondingly. We know from present experiments that hydrogen atoms become ionized, break up into unbound protons and electrons, at about 3000 degrees Kelvin. So at that time that must have been the temperature, and because the background radiation today is a thousand times cooler, space must have expanded by a factor of a thousand since then.

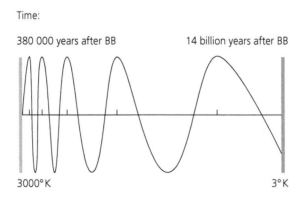

The stretching of the photon wavelength through the expansion of space

Nevertheless, the distribution of the photons contains otherwise all the information imprinted on them at the time of liberation, and that is the earliest direct information we can get about the beginning of our universe. Before that time, the world consisted of an interacting plasma of electrons, nuclei, and photons; after that, there were only electrically neutral atoms, so that the photons no longer had any partners with which they could interact. The formation time of atoms—the cosmologists call it the *time of last interaction* or *decoupling time*—this time is for us something like a temporal layer of clouds that stops our vision. All earlier happenings can be addressed only indirectly, and that is why today we have to search for the tiniest irregularities in the incredibly uniform distribution of the cosmic background radiation. These can perhaps give us at least some hints about what happened before, beyond the cloud surface.

Penzias and Wilson had run into the background radiation just by accident. They had been asked by the Bell Telephone Company to look at the possibility of microwave communication based on high-altitude balloons. In the course of their studies, they encountered a mysterious interference radiation, noise for which they could find no origin. This radiation was there day and night, and in all directions of the sky. Its origin became clear soon afterwards. A group of theorists at Princeton University had just predicted that there should be remnant radiation from the Big Bang, and as it turned out, that was the source of the apparent noise.

Penzias and Wilson only measured radiation of a fixed wavelength, but subsequent studies covered the entire spectrum. They found with unbelievable precision the distribution of black body radiation of a temperature of 2.725 degrees Kelvin. This is the radiation emitted by a container kept at a constant temperature of that value, and so it corresponds exactly to what one would expect if at an earlier time a perfectly uniform gas of photons had been liberated. In the following picture, we show the data taken by a special detector (COBE) installed on a space satellite. It is compared to the black body radiation mentioned earlier, although it's hard to see the comparison. We have here one of the rare cases in physics where the measuring error is considerably smaller than the thickness of the shown theoretical curve. With incredible accuracy, one observes throughout the universe black body radiation of a temperature of 2.725 ± 0.001 degrees Kelvin.

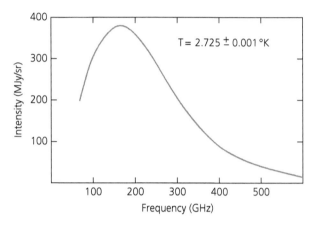

The spectrum of cosmic background radiation, theory, and COBE data

The measurement of this radiation was in fact not quite as simple as it seemed. On Earth, it would be like looking out of a steam-filled room through dirty windows. In particular, clouds and other gas effects have a considerable influence on the precision. That is why the stage of real precision data began only when space satellites became available, taking data far from any terrestrial interferences. The beginning was made by the Cosmic Background Explorer (COBE) toward the end of the 1990s; the results, including the spectrum just shown, brought the 2006 Nobel Prize in physics to the leader of the effort, George Smoot, and his collaborator John Mather. The next step was the Wilkinson Microwave Anisotropy Probe (WMAP); it was followed by the Planck detector, which in 2009 was brought into orbit by an Ariane rocket; we will shortly return to its latest results.

But all these measurements first had to be corrected for effects totally unrelated to the Big Bang, and in the spectrum shown previously they are already taken into account. These corrections sound a little like science fiction: we are in fact living on the

Spaceship Earth.

Our Earth orbits around the sun at some 30 km/s; the sun in turn moves around the center of our galaxy, the Milky Way, at 220 km/s; and even our galaxy is not stationary relative to the rest of the observable universe. The sum of these effects has us here on Earth moving at 390 km/s in a well-defined direction through space, through the container of the cosmic background radiation—the space, in which the entire matter of the observable universe is at rest.

What this means is well known from the world of sound waves. If we move toward the source of the sound, the wavelength is shortened; the pitch of the tone becomes higher. In the opposite case, moving away from the source, the wavelength grows; the pitch of the tone goes down. Physicists call this the Doppler effect, named after the Austrian physicist Christian Doppler.

And if we now travel on our spaceship Earth with a certain speed through the waves of background radiation, then the length of the waves in front of us is contracted, those in the back dilated. That implies an ultraviolet shift in the direction of travel and an infrared shift toward the back. These shifts are clearly observed; on the one hand, they tell us in which direction we are moving through space, but on the other, they have to be subtracted in order to obtain the correct spectrum of the radiation. In the following picture, this effect is schematically shown: here the colors are simply meant to indicate that the Earth is moving through space in the direction from the upper right to the lower left.

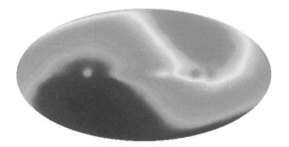

Doppler-shift of the cosmic background radiation
Reproduced with permission from DMR, COBE, NASA.

As mentioned, this effect has to be "subtracted" from the observed temperature distribution before one can draw any conclusions. A further effect is caused by clouds of gas that circulate in the plane of our galaxy. Just recently, some dramatic conclusions about the early universe had to be corrected, because the measured results were finally found to be due only to interstellar clouds of dust.

When all is taken into account, one obtains the following picture, recently taken by the Planck detector. We now indeed see deviations from a uniform temperature, but we have to keep in mind that the different colors indicate fluctuations of less than a thousandth of a percent: blue means a little cooler, red a little warmer. The information that the background radiation can give us about the universe before the time of last scattering is thus based on tiny variations. What can these tell us?

Distribution of the cosmic background radiation (from the Planck detector)

First, we should emphasize once more that this picture leads to an average uniform temperature of 2.725 degrees Kelvin, with that precision; and only then do we turn to the previously mentioned tiny deviations from the average. Why are photons coming from certain spots of the universe a little warmer or cooler?

Here we should remember that the inhomogeneous structure of the present universe, with stars and galaxies on the one hand, empty space on the other, must have come into being at some time. It's not possible that everything was completely uniform and homogeneous, and then suddenly the world became "clumpy." Even gravity cannot create clumps in a perfectly uniform world. For that reason, cosmologists have always insisted that one *must* find fluctuations in the temperature distribution of the cosmic background radiation, if one only looks carefully enough. And the improvement in observation techniques, from COBE to WMAP to Planck, did bring the required increase in precision.

So before the decoupling of the photons, in the hot plasma of nuclei, electrons, and photons, there were indeed certain, albeit tiny irregularities in the density of the medium. These must have first appeared in still earlier stages, as minute quantum fluctuations in the expanding primeval world, at the time of inflation. In the primeval medium, quantum theory tells us, there were again little bubbles of higher or lower density, similar to the bubble that originally led to our universe itself. Through inflation, the spatial extension of such fluctuations was dramatically increased, but at the same time, their deviations from the average were greatly reduced—everything was smoothed out, until today, only the tiny fractions of less than a thousandth percent remain. The macroscopic areas of higher density formed in this way later on gave gravity a chance to first create gaseous clouds, and still later form stars and galaxies out of these.

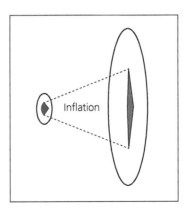

Expansion of a quantum fluctuation

When matter and photons became decoupled, photons coming from a region of denser matter had to overcome a stronger force of gravity than those from regions of lower density. The former were restrained, their wavelength was infrared-shifted, while the others had it easier and thus appeared faster, ultraviolet-shifted. The tiny irregularities appearing in the temperature distribution of the cosmic background radiation are therefore indeed witnesses of the fluctuations that were the seeds for the subsequent structure of the universe, the origin of stars and galaxies. What in the earliest times of the early universe was only a tiny spot of slightly higher density later on became the Milky Way, long after it had created the fluctuations in the spectrum of the cosmic background radiation.

But this radiation has still more to tell us. In the plasma, the attraction of gravity and the expanding pressure of the medium compete with each other. An area of higher density contracts under gravity; but this increases the temperature, so that the radiation pressure increases as well, and eventually the bubble expands again. Such a process, a temporal variation and contraction of a medium, is in fact quite familiar to us: it happens in sound waves. So in a sense, at the time of last scattering we encounter not only the light, but also

The sound of the Big Bang.

The density oscillations in the plasma act similarly to an underwater bellows that is contracted and released in a certain rhythm. The resulting pressure pulses create sound waves that propagate through the water. Whales and dolphins use this effect for communication, and our sound waves in air are of this nature. In this way, the small density irregularities created by quantum fluctuations cause the early universe to "sing." The spatial regions formed at the end of inflation form "sound boxes," whose size is determined by the speed of sound waves in the plasma medium. In other words, the size of the sound box is the region accessible by sound waves up to the time of decoupling. The speed of sound waves in a relativistic medium is about 57% of the speed of light, and so the volume thus created has a spatial radius of some 200,000 light years and a lifetime of 380,000 years until decoupling sets in. The modification of density modifies gravity effects, and as a result the spectrum of the photons is slightly modified by this hill-and-valley structure. Because at the time of last scattering, matter and photons become decoupled, the bellows stop working at this point—the song is frozen. And the challenge for satellites is then clear: measure sufficiently precisely to identify the frozen melody.

The following picture shows what this requires. Observers have to take their bearings in the radiation sky at such a small angle that they cover exactly one sound box and not more, because otherwise other sounds come in and create a cacophony. As mentioned, the sound box had a size of about 200,000 light years at decoupling time; the subsequent spatial expansion increased this by a factor of 1000, so today we get 200 million light years. For the observer, that means an angle α of a little more than one degree. Up to this angle, one observes the emission of the primeval sound box; for larger angles, the result becomes an average over several emitters and thus approaches eventually an equidistribution. Only for the noted small angle can we hope to hear a pure sound.

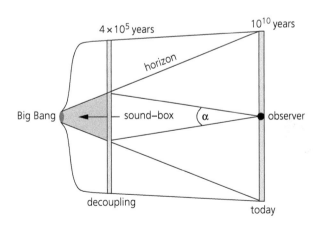

The sound box of the Big Bang

The wavelength of the sounds produced in this way is determined by the velocity of sound in the given medium and by the size of the sound box, which is here given by the length of time available. Using the values quoted earlier, we get a wavelength of about 200,000 light years for the fundamental tone, measured at the time of decoupling. As noted, since then the universe has expanded by a factor of 1000, and therefore the waves of the cosmic sound have increased accordingly. As a result, the wavelength of the fundamental tone comes for us to 200 million light years. In addition to the fundamental tone, there are harmonics, overtones, of higher frequencies, integral multiples of the fundamental tone frequency. In the following picture, this is illustrated schematically for a closed flute.

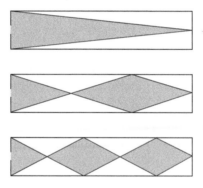

The fundamental tone and the following two harmonics (top to bottom)

All these waves thus created a topography, a landscape of hills and valleys, through which the photons had to work their way out into our world. Because of this work, their spectrum was shifted up or down and thus led to the spotty carpet of cosmic radiation that we observe, if we look carefully enough.

To catch the fundamental tone, one has to observe radiation in an angle of one degree, as we have noted. To get the higher harmonics, one has to observe at correspondingly smaller angles, at finer resolution. What emerges in this way as a function of the angle of observation is illustrated in the following picture. The apparent regularity of the harmonics is, however, an idealization, because in reality, the remnant medium present in the interstellar space leads to an effective dampening.

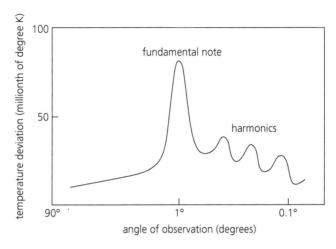

Variation of the temperature of the cosmic background radiation as a function of the observation angle

What can we learn from such measurements? An essential question can be answered with the help of the first maximum: what is the form of the space in which all this takes place? We remember that a long time ago, many people pictured the Earth as a disk, a plane—even though in ancient Greece doubts were already being expressed about such a scenario. We now want to repeat this question in a very much larger framework.

The form of the space

in which our universe "lives"—the form of the stage on which everything takes place—is an issue that has to be clarified. The most familiar form is of course that encountered in our normal world: a flat space in which two parallels never meet and in which the angles of a triangle add up to 180 degrees. We find it normal to picture all space like this; but then, our ancestors also found the Earth as a disk perfectly normal. We get an alternative if we consider the surface of a sphere as the underlying space. Now all parallel lines emerging from the equator at right angles meet at the North Pole—they are our lines of longitude. And the angles of an equilateral triangle, with the equator as baseline, add up to 270 degrees.

Such a world can be extended to three (or more) dimensions, even though everything then becomes a bit more complicated. But we can readily distinguish between the two possibilities, flat or spherical space. If two explorers start out at the North Pole, with an angle of 90 degrees between their lines of travel, then they arrive at the equator after some 10,000 kilometers, separated by a distance of again 10,000 km. If the Earth were flat, their distance of separation would be much greater, some 14,000 km. In other words, in a spherical world distances "shrink"—the 14,000 km becomes 10,000 km.

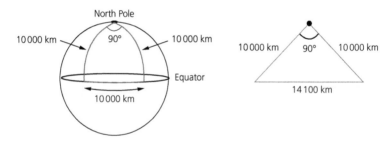

Distances on the surface of a sphere (left) and on a flat plane (right)

Fortunately, we can stick to our well-known view of space as flat. If we know how large the sound box was at the time of decoupling, then we find in flat space (for two dimensions we would have a planar triangle) indeed the angle mentioned earlier, about one degree, for the first maximum. For a spherical world, the angle would become smaller, in two dimensions by a third. The observations show that that is not the case—our universe is flat, just as our closer surroundings are.

At this point, it is perhaps useful to distinguish two features: the form of space and its rate of expansion; we will deal in more detail with the latter in Chapter 8. For the form of space, we have three different types: the two just mentioned, flat or spherical, and in addition hyperbolic (saddle-shaped). If we project the behavior of triangles onto a plane for each of these three cases, we get the following picture.

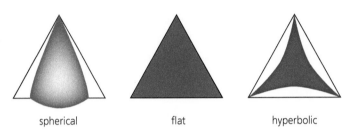

spherical   flat   hyperbolic

Triangle structures for different forms of space

While in flat space, the three angles add up to 180 degrees, their sum is larger in a spherical world, less in a hyperbolic.

The stage for the actual cosmic background radiation is, however, not really empty: our universe contains assorted bits and pieces, and that, as we know from Einstein, has an effect on the spatial structure. Obviously there is the "visible" matter known to us, planets, stars, galaxies. In addition, there is a further participant, one that is quite problematic for present-day astrophysicists. Galaxies such as the Milky Way are bound together by the force of gravity between the participating stellar constituents; gravity determines the shape and size of the galaxy just as the sun and the planets determine the size of our solar system. For the Milky Way, just as for the solar system, the motion of all constituents has to be in accord with the gravitational forces active in the system. That, however, does not work out: to bind a galaxy such as the Milky Way and to arrive at the motion found for its stars, one needs more matter than is visible, very much more; in Chapter 7, we will return to this issue in more detail.

So once again there has to be an invisible player in the game, *dark matter*, not to be confused with dark energy. Dark matter surrounds galaxies, while dark energy is uniformly distributed throughout the

entire universe. The amount of dark matter needed for the shape
and behavior of galaxies is many times that of the visible—it must
be there, but it cannot be seen; it interacts with our world only
through gravitation. How something like that fits into our framework of physics remains so far an open question; not even a black
hole can do it, because dark matter has to be distributed over a very
large region, whereas black holes are extremely dense and concentrated. Whatever the solution, the universe must contain altogether
very much more matter than we can see. And the total amount,
dark and visible, contributes to the effect of gravity, to the deformation of space. To counterbalance that, we need the often-mentioned
dark or space energy, which has the opposite effect. Based on today's
observations, this space energy must add up to about 75% of the total
energy of our universe: we have 75% dark space energy, 20% dark
matter and 5% visible matter, as illustrated in the following picture.

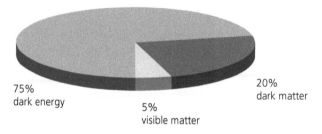

The composition of the universe

Let's reiterate the underlying logic. Visible matter we can measure. The structure of the Milky Way requires four times more
dark matter, as we shall see, in order to arrive at the force of gravity needed for shape and binding of the galaxy. The position of the
first maximum in the spectrum of the cosmic background radiation
shows that the space of the present universe as a whole is flat. That
requires the additional 75% dark energy, whose force of expansion
compensates the gravitational attraction of dark and visible matter,
leading to a flat space. As mentioned, we'll return to all this in more
detail in Chapter 7.

The further investigation of the cosmic background radiation is today a major topic of research. Here we want to address a subject recently discussed quite often, the polarization of the radiation. Light is an electromagnetic wave, whose amplitude "oscillates" orthogonal to its direction of motion; if we stand at a given point through which it passes, we experience an alternation of peaks and valleys. Normal light consists of a superposition of such waves, oscillating in all directions at right angles to the line of motion. Such light is referred to as *unpolarized*. If a beam of such light is passed through a slit allowing only oscillations in one direction, then the light on the other side oscillates only in the direction of the slit; the other components are stopped: the light is now *polarized.*

In the following picture, we show for simplicity only two waves oscillating at right angles to each other, of which one is stopped by the orientation of the slit. The stopping of the other components not along the line of the slit of course reduces the intensity of the light, and therefore this effect is often used in sunglasses. The light reflected from a surface of water is largely polarized to oscillate in the plane of the water surface; sunglasses made of material with slits orthogonal to the water thus stop much of the incident sunlight . . . unless you turn your head by 90 degrees.

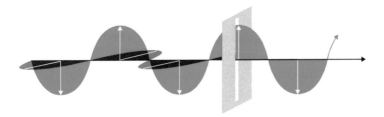

Polarization of a light beam

What can we say about the polarization of the cosmic background radiation? The light reflected from a surface of water is not, as one might think, a "redirected" beam; instead, it is a completely new beam. The incident photon hits an electron in the water surface and gives a kick. As a result, such kicked electrons in turn emit "secondary" photons, and the photons created in this way form the "reflected" light we observe. The motion of the electrons triggered in this way occurs largely on the surface of the water, and this causes the newly created photons to oscillate in this plane.

The origin of the photons in the cosmic background radiation is quite similar. These photons are also "reflected": they were created in the last collision of plasma photons with electrons. However, the electrons now are not in a plane, and the plasma photons come from all sides. As long as all photons are equally strong, there is no preferred direction, and the resulting radiation is unpolarized. But if the intensity of the incident photons depends on their direction of motion, then the emitted photons will transmit the information of this direction through polarization.

As a result, the observation of the polarization of the cosmic background radiation can provide information about eventual bumps or irregularities in the universe just before the threshold of last scattering. Such *anisotropies* can arise through fluctuations in density or temperature, but they can also be triggered as an aftereffect of gravitational waves created at the time of inflation. Even at the much later time of last scattering, such waves can still lead to remnant turbulences. The polarization formed as a result of such gravitational waves is quite different from that due to density or temperature fluctuations, so that their observation would provide a first hint for the existence of inflationary waves. The polarization due to density of temperature fluctuations is symmetric, pointing toward the center of the irregularity. If there is in addition a spatial wave surviving from the time of inflation, then this wave will provide direction and thus destroy the symmetry. The effect of such a gravitational wave is expected to be very weak and hence very difficult to measure.

It is therefore no surprise that the announcement made two years ago by the American research team BICEP led to great excitement: they thought they had found the first evidence for asymmetric polarization due to gravitational waves. But, as in many other cases, unfortunately the champagne corks popped here too early. Interstellar dust clouds can lead to a similar effect, and in the direction

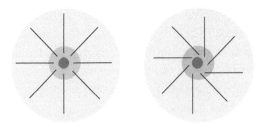

Polarization scheme for a density fluctuation (left) and the modification
due to a gravitational wave (right)

of the sky studied by BICEP there were such clouds. More extensive
studies in different directions of the sky, carried out with the European Planck detector, showed that when the clouds were absent, so
was the effect seen in the polarization.

Nevertheless, the great interest in gravitational waves persists.
Until now, electromagnetic waves are our only line of communication with times long gone. But neither dark matter nor dark energy
can experience electromagnetic interactions—they can only communicate through gravity. And just as electromagnetic interactions
lead to the emission of light waves, gravitational interactions must
produce gravitational waves—so predicts Einstein's general theory
of relativity. What is so fascinating is that these waves give us direct
information about the deformation of space and time; what is problematic is that these spacetime waves are very weak and therefore
difficult to detect.

Let us try to construct a detector for this purpose. We start with
a pipe several kilometers long and evacuated, to eliminate all other
possible disturbances. On one end of this pipe we install a light
source, on the other a mirror. Now we send out a light signal and
measure how long it takes until the reflected signal returns. If during the measurement a gravitational wave passes the site, then the
path of the light signal is sometimes longer, sometimes shorter, and
that should show up in the time of travel of the signal.

It is easy to imagine that the time measurement for such an
arrangement is hardly possible on a terrestrial scale: for a length
of ten kilometers, light needs thirty-thousandth of a second, and

to measure tiny deviations of this is well beyond the possibilities of most watches. To avoid this difficulty, the experimentalists use a setup in which two such pipes come together at a right angle. The light beams meeting at this corner point create interference effects, and if the lengths of their paths are contracted or dilated, these interference effects are modified. With the help of such a setup, it indeed becomes possible to measure wave deformations of space.

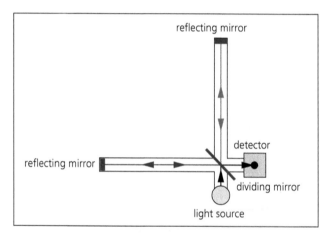

Gravitational wave detector

The light beam is split by the dividing mirror and sent into two arms; at the end of each arm, the corresponding beam is reflected. The reflected beams, when they meet, create an interference pattern, which is measured in the detector. Gravitational waves modify the effective arm lengths, and thus change the interference pattern.

In February 2016, a team of scientists from the USA and Europe (LIGO) used two such detectors well separated (one in the state of Washington and one in Louisiana in the USA), and they received signals—according to the analysis—due to gravitational waves. These waves are not remnants of Big Bang inflation; they were created in the collision of two massive black holes some 1.3 billion years ago and have arrived here only now. If these measurements are confirmed—and after the BICEP affair one has become even more careful—then this direct proof of gravitational waves would constitute an absolute breakthrough.

Information about the collision of the two black holes would only be the start of a new way of information transmission from distant times. As we indicated in connection with the BICEP experiment, the inflation at the birth of our universe must have created gravitational waves that even today cause spacetime to swing, albeit very weakly. However, from what has been said, it is clear that terrestrial experiments soon reach their limits—we have to let light travel over much larger distances in order to see clear modifications.

The hopes and expectations of scientists therefore rest on the new eLISA project, a joint effort of several European laboratories. It is based on an interferometer situated on one satellite in space, reflectors on others, so that the effective length of the "pipes" is increased from ten to a million kilometers. If all goes according to plan, the measurement and investigation of gravitational waves will be carried out with this new tool—perhaps as fruitful as cosmic microwave radiation has so far proven to be.

In closing this chapter, we want to briefly mention a further development of the universe. Nucleons and electrons are now bound to electrically neutral entities, so that cosmic microwave radiation can proceed unhindered through space. The small fluctuations in density of matter now begin to show consequences. As long as radiation and matter were in interaction, radiation pressure could break up eventual clumps of matter created by gravity. But now radiation is out of the game, since matter has become electrically neutral. Now denser clouds of gas can form, which through gravity attract other such clouds and thus become larger and still denser. With time, these "proto-stars" will become real stars: with increasing density, the kinetic energy of the nucleons became ever larger, until a point was reached at which the collision of four nucleons would form a helium nucleus: nuclear fusion had started. The energy liberated in this process is sent as radiation into space: the star has started to shine. This

Star rise

began some 500 million years after the Big Bang and thus ended an era which the cosmologists call the dark ages, since between the last scattering and the creation of stars nothing was shining, apart from the cosmic microwave radiation, of course. But now the lights go on in the universe, stars light up the sky, darkness is over.

The appearance of the stars leads us to our next big question: how could this universe, this hot atomic gas, uniform up to a millionth of a percent, produce all the stellar structure of our present world? We recall that there are laws of thermodynamics claiming that disorder, lack of structure will always increase in the course of time. What does that mean, and how, nevertheless, could stars and galaxies appear on the scene?

6

Structure and Form

The whole is more than the sum of its parts.
ARISTOTLE, *METAPHYSICS*

Structure and form surround us wherever we look. According to the Bible, in the beginning, "the Earth was without form and void, and darkness was over the face of the deep." The physics of the newly born universe assumes the absence of all forms and scales; just after the Big Bang, the universe consisted of pure, concentrated energy, without the least structure. There was not even nothing, no empty space, and there was not the slightest indication of all the complexity that would eventually arise from this simple primordial state. Nevertheless, the seeds of the coming diversity must already have been present in the force fields making up the uniform primordial medium. Although there was no way to see it, the structureless world must have already contained in a latent form the seeds for the structures later to emerge. The appearance of structure out of many simple identical components is today found in many areas of science, and is in fact generally denoted as *emergence*, as collective self-organized formation. And not only structures are emergent; there also exist emergent observables of all kinds: temperature, density, pressure, to name just a few. A single atom or molecule does not have a temperature; such quantities characterize the collective behavior of many individual components.

Emergence is familiar in our daily life: out of an apparent uniformity, in the absence of any form, suddenly specific formations can appear. We have discussed the states of water. Steam consists of

very weakly interacting molecules; it is a gas of almost free particles, the same in all directions and completely without structure. There is no way to tell that with decreasing temperature such a system will first suddenly liquefy and then eventually turn into ice with its crystal structure. However, the basis for such transitions must have been imprinted in the form of the interaction of the gas, like inheritance in genes. The complexity arises *spontaneously*, without any external action; in a latent form, it must therefore have been present all along. In this way, the laws governing the newly born universe, completely without any structure, just after the Big Bang, before the existence of gravity or electromagnetism, of nuclei, atoms, or molecules—these laws must have already contained the prerequisites even for the formation of an ice crystal or a snowflake.

Wherever we look today, we see a world of infinite diversity, of structures, colors and sounds, of processes and events. As far as we know, even animals are aware of such a diversity. We humans, in contrast to them, can reflect and question. An African friend summarized our drive to understand in the words "We used to see the sun, the stars, the clouds, the mountains, the animals and the trees, and we enjoyed what we saw. Today we want more." What is this *more*? What do we want to understand? Is the symmetry of the figures that we see a reason for our enjoyment? Or is it the order, the underlying pattern that we recognize when the seasons of the year keep reappearing? When did humans first realize that the moon provides them with a calendar, that the flow pattern of the tides changes, but periodically returns? At some time long ago, it became clear that diversity does not mean chaos, that there was some inner order, some rules governing the occurrence, constructed such as to give rise to a spatial structure and to a direction of time.

In the beginning, diversity also led to the desire to order things such that combinations "made sense," were based on common features. What are the possible forms of matter? Is it possible to reduce the observed states to a smaller number of basic forms? In the fifth century BC, the Greek philosopher Empedocles defined four forms, four *elements*:

Earth, Water, Air, and Fire

The Sicilian earth, the water of the Mediterranean, the winds along the coast, and the fire of the volcanoes suggested these as natural forms. Not surprisingly, similar considerations also appeared in the early Buddhist-Hindu thinking. And in both cases, the question came up as to what the stage for these forms might be: where does matter appear, where are its building blocks put together? In Greek thinking, this led to a fifth form, the *quintessence*, empty space.

For the Indians, the "Vakasha," the nothing, played a similar role. The different forms of matter stage their appearance in front of the background of empty space. These concepts have not really changed much in the course of the past two thousand years: even today, solids, liquids, gases, and plasmas are the basic forms of matter. And if we heat a solid at constant pressure, matter passes with increasing temperature through these four states; a plasma does so for decreasing temperature. But the empty space of today is no longer just *nothing*. On very large as well as on very small scales, our concept of the vacuum has become more complicated.

How did these forms appear, and why—why just these? Once Albert Einstein asked, "When God made the world, did he have a choice?" That seems to be a philosophical question, or at least a metaphysical one. But present-day physics can contribute to the answer, and that is what we want to show here. Our world was not always as it is now, and by investigating how it got to be this way, we can learn to understand many things. The path from the Big Bang to today evidently was one from a simple, uniform world to all the complexity that surrounds us now. At one point, heaven and earth were separated, light and dark, wet and dry, and much more. How could that happen, how could structures develop? How could the hot uniform clouds of gas just after the Big Bang lead to a solar system of circling planets? Is it possible that a completely disordered world turned into one with so much structure?

The direction of the evolution of systems in our present world is codified, written down as one of the immovable laws of physics, as a basic law of thermodynamics. It is the *thermodynamic arrow* that points toward expansion, mixing, breaking, aging. A broken glass will never recombine again on its own, a scrambled egg will never again become a whole egg, nor will ashes turn into coal. There is a direction that so it seems, leads from structure to its absence, from order to disorder.

The course of evolution

is thus specified in today's physics, and it defines a direction for the progression of all physical phenomena. It took physicists quite a while to reach this conclusion, or—to be more honest—to admit it. The equations of mechanics or of electrodynamics don't define a temporal direction—everything can evolve forward as well as backward in time. However, if we look at systems consisting of very many particles, that fact is not of much help.

To see why this is the case, it is useful to consider an experiment carried out around 1850 by the British physicist James Prescott Joule. Joule was the son of a brewer, and he later took over his father's brewery; hence his interest in things like temperature and pressure was not purely academic. In his experiment, he started with a container separated into two compartments, with the help of a dividing wall, and isolated from the outside world as much as possible. In one compartment, there was a gas; in the other, nothing, a vacuum. If Joule now removed the dividing wall, the gas expanded until both compartments were equally filled. And he could wait as long as he wanted—the gas in the previously empty compartment never returned to its initial volume. The expansion of the gas was certainly a physical process, but it never ran backward: the direction of evolution was fixed; the evolution was *irreversible*. In principle, the gas was free to return from time to time to its starting compartment, like a liberated animal to its cage—but that never happened.

Other features of the process are also quite interesting. Because the whole setup was isolated from the outside, the total energy of the gas was conserved in the expansion: the number of molecules and their momenta were unchanged, and so also the temperature remained the same. But the available volume was now larger, so the density of the gas had been reduced. Also, the pressure, the total energy of impact on the container walls, had become smaller: per square centimeter, there were now fewer impinging molecules, because the total area had grown, but the number of molecules had not.

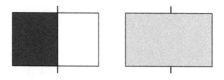

The Joule experiment

The understanding of the situation required an enlargement of the conceptual world of physics. When you study a gas in a container, you measure its temperature and pressure. These quantities, however, do not arise from one or two molecules, whose paths are forward-backward, symmetric in time; they are the result of the collective effort of all 10^{23} molecules of the gas. A few molecules might once in a while find their way back into the initial container, but that all the weakly connected particles *simultaneously* turn around: that just doesn't happen. If only a single particle on its way back is through some interaction deterred from its path, then the entire return collapses. The investigation of individual tracks thus becomes meaningless. What you can determine is the average energy of a particle, for which temperature is a measure; the average number of particles per unit volume, that is, the density; and the average impact energy on the walls of the container, the pressure. The reversibility of individual molecular fates becomes insignificant, their paths unimportant; they become submerged in the mass of all, and for the evolution of that, there is a temporal direction.

It thus became clear that the description of the collective behavior of very many particles requires a new postulate. Something had to be added to the previous laws of physics, which allowed time reversal. Mainly through the work of Ludwig Boltzmann in Vienna and John Willard Gibbs at Yale University in the USA, these thoughts led, toward the end of the nineteenth century, to a new paradigm, *statistical physics*. Imagine a gigantic catalog containing all the possible configurations of all the 10^{23} molecules allowed when the overall parameters of the system are fixed, that is, its total energy, its volume, and the number of molecules it contains. We call these overall parameters the *macrostate* of the system.

A specific configuration, sometimes called a *microstate*, consists of all positions and momenta of all particles possible at a given time for a given macrostate; here we assume classical statistical physics. The basic postulate of statistical physics now says that a system in thermal equilibrium finds itself in principle, a priori, with equal probability in any one of the multitude of microstates, provided there is no interference from the outside. And that means in the evolution of the system from one microstate to the next, in the buzzing swarms of particles, the most common states win—unusual configurations are almost never reached, and thus become effectively eliminated. We shall see shortly that in the Joule experiment, the number of configurations possible for a gas equidistributed over both compartments is so incredibly much larger than that for the gas in the initial compartment only, that this indeed rules out any return.

To illustrate what happens in such cases, let us look at a box with nine compartments. If we now take four identical balls and distribute these into the compartments, one per compartment, then there are in total 126 different configurations. In the following picture we show three of them, and in Appendix 1 we do the corresponding counting. What we want to emphasize here is that there is exactly one configuration in which each ball is in a corner. If we assign to all configurations the same probability, then the chance for such a corner arrangement is 1:126. In other words, such very "ordered" microstates are rather unlikely.

Configurations of four balls in nine compartments

Next we want to see what effect a doubling of the size of the box has, giving it 18 compartments. Now there are 2060 configurations for the four balls, that is, almost 20 times more than in the previous case. And this relation between starting size and doubled size increases more and more with the number of balls. If we start with nine balls, we fill all the compartments, so that there is just one configuration. The doubled volume, however, now gives 48,620 possibilities for the nine balls; the chance to return to the starting position is 1:48,620.

 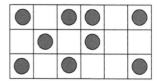

Nine balls in nine and in eighteen compartments

That ratio holds for only nine balls. For a box with more compartments and a larger number of balls, the chance for a return to the starting configuration decreases more and more. And if we more than double the size of the volume, it drops even more. For an x-fold volume growth and n balls, the number of configurations increases as x^n (again see Appendix 1). In the case of a normal gas, with some 10^{23} molecules, the opening of the new compartment leads to the incredible increase in the number of configurations mentioned earlier. If we assign to all configurations the same probability, then the chance to find the gas again in the initial compartment is reduced by the astronomic factor $x^{100\ldots000}$, an x to the power of a one followed by 23 zeros. In comparison to this, winning the lotto (six out of 49) is an easy task: for 36 numbers, there are "only" about 14 million configurations of six different numbers between 1 and 49.

The fundamental quantity for the behavior of systems in statistical physics is thus the number of possible configurations, of *microstates*, for a given total or *macrostate*. The measure for this quantity is the

Entropy.

If we know the total energy E of the system, the number N of molecules it contains, and the volume V of the container, the macrostate is fixed and we can calculate the number $W(E,N,V)$ of possible microstates. The entropy $S(E,N,V)$ is then defined by the famous formula

Tomb of Ludwig Boltzmann in Vienna
Photo courtesy of Oesterreichische Zentralbibliothek
fuer Physik, Vienna, Austria

inscribed on the tombstone of Ludwig Boltzmann in a Vienna cemetery. The proportionality factor k is named after him and provides the connection between dynamics and thermodynamics. The Boltzmann constant k, the gravitation constant G, the speed of light c, and the Planck constant h then give the four fundamental constants of physics, for thermodynamics, gravitation, relativity theory, and quantum theory, respectively. For the definition of entropy, one uses the *logarithm* (log); this has two helpful aspects. The number of microstates is immense, and so the logarithm is more suitable, because it just gives the number of powers of ten. To illustrate: the logarithm of a billion, $1,000,000,000 = 10^9$, is simply 9. Furthermore, the number W of microstates for a system consisting of two parts is the product of the two numbers, W_1 and W_2, $W = W_1 \times W_2$. Because the logarithm of a product is the sum of the logarithms of the factors, the entropy of the total system becomes simply the sum of the subsystem entropies, $S = S_1 + S_2$. For a gas, as in the Joule experiment, the number of microstates is determined by the available volume, and so the overall entropy becomes the sum of the partial volumes. The difference in entropy before and after removal of the divider is thus determined by the resulting increase in volume.

The Joule experiment shows another, essential feature: a system left to its own, isolated from the outside world, always develops so as to maximize the number of microstates allowed, to maximize its entropy. And once it has reached that macrostate, it will remain there forever, barring outside actions. Before the opening the divider, that was the case: the system was in a state of maximal entropy for the initial volume; it was in thermal equilibrium. The opening of the divider suddenly increased the available volume, so that the system now was no longer in equilibrium, because its actual entropy at that moment was much less than that for the larger volume. By streaming into the new volume, the gas increased its entropy, until it finally reached the maximum value for a gas at fixed temperature in the larger volume. The system was now (again) in thermal equilibrium.

The two basic laws of thermodynamics summarize these considerations. The first says that the *energy* of the total system remains constant, is *conserved*: without actions from the outside, it can neither increase nor decrease. The second law of thermodynamics states that the *entropy* of an isolated system *never decreases*, and that in equilibrium it attains its largest possible value, its maximum. After that, nothing changes anymore: after all, equilibrium means that everything remains as it is, provided nothing interferes from the outside.

It seems worthwhile to elaborate a little further the connection between entropy and structure. To write down the previous sentence, you need 81 letters. The English alphabet consists of 26 letters. So there are 26^{81} ways of arranging the 81 letters—in other words, there are 26^{81} microstates, more than 10^{114}. Only one of all these produces the sentence in the given form. Structure, and here information, corresponds to a very low entropy. Conversely, recognizable structures disappear with increasing entropy—the transfer of information ends. News can be transmitted by Morse code only because one sends a very specific sequence of all the possible short-long combinations. The higher the entropy, the lower the information contained in the system.

The concept of entropy is in fact older than its explanation in terms of the motion of atoms or molecules of the medium, as we have given it here. The French physicist and engineer Sadi Carnot had noticed around the beginning of the nineteenth century that heat always flows from hot to cold, never the other way around. He was presumably not the first to see that, but he recognized this as one of the basic features of what was to become thermodynamics. Just as the atoms in Joule's container would never stream back into their initial compartment, a pot of water would never spontaneously freeze and thereby heat up the room. Twenty years later, the German physicist Rudolf Clausius used that to define the concept of entropy, thus establishing the fundamental quantity for a theory of heat. The subsequent development of statistical physics then provided the explanation in terms of the atomic structure of matter.

Up to our time, the second law of thermodynamics, *the entropy of an isolated system never decreases*, is perhaps the most profound statement of physics. In a science fiction world, in which it is not valid, cold water can become hot without heating, the old can become young again, the dead arise, and much more. There are few statements of physics so directly connected to the experiences of our daily lives, few that form such a fundamental basis of all our knowledge about the course of events. The famous English physicist and astronomer Sir Arthur Eddington formulated that as advice to his colleagues:

If someone points out to you that your pet theory of the universe is in disagreement with Maxwell's equations, then so much the worse for Maxwell's equations. If it is found to be contradicted by observation—well, these experimentalists do bungle things sometimes. But if your theory is found to be against the second law of thermodynamics, I can give you no hope; there is nothing for it but to collapse in deepest humiliation.

As we shall see, that is no empty threat.

The concepts presented so far allow us to understand the steps, the direction in the evolution of matter. A glass that is intact has that one state, corresponding to minimal entropy. The harder it falls or is thrown, the more pieces there will be, so that the final entropy, the number of produced pieces, depends on the energy put into its breaking. We can further increase this energy by heating. The fragments now melt and form a liquid of glass molecules, which eventually will evaporate and form a gas. The addition of still more energy will first break up the molecules into their atomic constituents and then even the atoms into nuclei and electrons, leading to a plasma of electrically charged constituents and thus to still more microstates.

Based on the concepts of conventional statistical mechanics, we thus expect that a system initially not in equilibrium—the falling glass, its melting fragments, the gas in the initial compartment of Joule's experiment—will rapidly evolve, so that its entropy, too low for the given conditions, will increase to attain the maximum value in each case: many fragments, molten fragments, the gas in the entire volume. This evolution always proceeds from a state of low entropy ("structure") to an equilibrium state of maximal entropy ("uniformity"). That is the course of events prescribed by "normal" physics. Fluctuations remain possible—some corner of the world can accidentally be a little more structured—but on the whole, the direction is set: larger entropy, more disorder, less structure. On a warm spring day, a winter scene with snow, snowmen, and icicles is no longer in equilibrium: everything first melts and then evaporates, until all these structures have become uniform water vapor.

If then in the development (evolution) of all systems entropy can never decrease, how can we understand that the Big Bang, starting from a structureless hot gas, has led in the long run to the diversity of our present world, with galaxies, crystals, and snowflakes? In other words, how can the second law of thermodynamics, insisting that the entropy of an isolated system can never increase, that order leads to disorder, be reconciled with the apparent formation of

Structure in the universe?

That problem has occupied physicists and cosmologists for quite some time, and even today there is by no means universal agreement. Here we only want to present one picture of how it could have been. The Big Bang lies 14 billion years back, and so it is quite understandable that the picture we have of the earlier stages of the universe still shows some speculative features. But one thing we should always keep in mind: Eddington's warning words.

For quite a while, the apparent dilemma between structure formation in the universe and the second law of thermodynamics was resolved in a way that predicted a rather unpleasant end of the world. These considerations go back to Lord Kelvin, Hermann von Helmholtz, and others around the middle of the nineteenth century. They assumed that the universe was initially in an ordered state of low entropy (prepared that way by a Creator?) and subsequently proceeded to evolve slowly but surely into an ever more uniform state of maximal entropy. The predicted result is the "heat death" of the universe: the whole world would end as a uniform thermal medium, a gas without any structure and any energy-consuming processes. In the final chapter of this book, we shall return to these ideas. Such an end is still not fully excluded, but if at all, it will happen only in the very long run.

Today we see the development up to the present quite differently. There are basically two aspects that decisively influenced the evolution of the universe after the Big Bang. On the one hand, the universe has been expanding ever since then and continues to do so.

We therefore have to check if such an expansion in fact ever allows a thermal equilibrium to be attained. It takes a certain time to reach equilibrium, and if the expansion is fast enough, that time is not available. On the other hand, it is not unambiguously clear how a state of maximal entropy has to look. In our world of snowmen and icicles, raising the temperature above the freezing point evidently first produces water and eventually, given enough heat, water vapor—a uniform, disordered system of water molecules. Here the form of the intermolecular forces plays a decisive role, and the nature of the force between the constituents of matter is important in other cases as well. For a medium of non-interacting or only weakly interacting particles, the state of maximum entropy is always a disordered, structureless gas. But if the interaction becomes stronger—for example, at lower temperatures—maximum entropy can mean crystalline ice, with a well-defined crystal structure. Something similar happens if the particles carry electric charges, positive and negative, in equal numbers. At high temperatures and densities, the state of highest entropy is again a uniform disordered plasma. For dilute or low temperature media, however, a positive and a negative particle combine to form an electrically neutral entity, an atom; that allows both to escape from further electromagnetic interactions. Now the state of highest entropy is a disordered gas of such atoms. Later we will see that for stellar clouds in the cosmos, gravitation plays a still stronger role. Here we shall first consider in more detail each of the two factors we have mentioned as crucial in the evolution of the universe: expansion and interaction.

To understand the role of expansion a little better, we return to the Joule experiment and consider what happens when the dividing wall is removed. In the moment of the opening, the gas is still completely in its initial compartment, where it was in thermal equilibrium, that is, in a state of maximum entropy. The opening suddenly provides access to a new and empty additional compartment. The gas begins to flow rapidly into this compartment and to fill up the empty volume. After a certain "relaxation time," both compartments are filled up, and as soon as the entropy has reached its new larger value, corresponding to the larger volume, thermal equilibrium is once more attained. In the time between the opening and the re-establishment of maximum entropy, during the relaxation time, the system is not in equilibrium; its entropy is larger than before the removal of the divider, but less than the allowed maximum. During the entire process, the entropy is increasing, but it does not reach its new maximum immediately; that happens only after the relaxation time is over, and from now on, everything remains stable. The temporal behavior of the entropy is illustrated in the following picture.

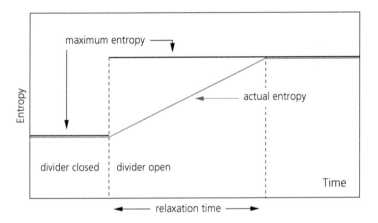

The temporal evolution of the Joule experiment

During the relaxation time, the medium is not without structure—on the one hand, it consists of partially empty, partially gas-filled regions; and on the other hand, the gas-filled region is expanding, moving toward the opposite container wall. Shortly after the opening of the divider, the molecules that have entered the new volume are not moving around uniformly in all directions— they are moving mainly toward the opposite wall, the faster ones in front, the slower further back. So there clearly is some kind of "order." In the course of the relaxation time, this order disappears and finally all molecules are equidistributed over the entire volume;

the entropy has reached its new maximum value, and the system is in equilibrium once more. An intermediate state, in which equilibrium is not yet attained, is shown in the following drawing.

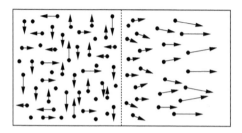

Molecules in the Joule experiment shortly after the removal of the divider

What's happening there has in recent years led to some remarkable considerations. It seems as if some mystic "entropic" power is driving the molecules from left to right. A single molecule would simply fly around—it would be in either section of the container with equal probability. It would not experience this power, which therefore must be a collective, multi-particle effect. In the following chapters, we shall return to such *emergent forces*, which arise only from the interplay of many individual constituents.

In the Joule experiment, the increase of the maximum possible entropy was achieved by the removal of the divider, which suddenly created a larger accessible volume. But we can also do this more gradually, by pulling out a piston (for simplicity without any friction), and thereby increasing the volume. Here it is crucial that we pull and not have the gas pressure push: that would cause the gas to do work and reduce its temperature. What we have in mind is achieved if the piston is very heavy, so that the gas pressure cannot move it.

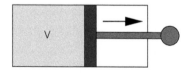

The continuous Joule experiment

In such an experiment there are two rates of change: the relaxation time, which the disturbed gas needs to return to equilibrium, and the rate at which the volume is increased by pulling out the piston. If we move the piston very slowly (*adiabatically*, in physics terminology), the gas can adjust to the changing conditions: it remains always in equilibrium; the increasing entropy always stays at its (increasing) maximum. But if we retract the piston very rapidly, the gas cannot keep up: it falls more and more out of equilibrium; it becomes ever more "ordered," the molecules frantically try to reach the disappearing opposite wall. Instead of a uniform medium, we now have oriented flow, attempting to reach the retracting piston.

In other words, the crucial competing phenomena are

Expansion vs. relaxation time.

If the accessible volume increases faster than the relaxation time needed for equilibration, then the maximum entropy can never be attained. On the contrary, the difference between actual and maximum entropy continues to increase with time—the system shows more and more structure. This phenomenon—the role of an expanding volume in the context of the second law of thermodynamics—was proposed around 1975 by David Layzer of Harvard University as a possible source of structure formation in the universe. If you increase the available volume for a gas in equilibrium, then the rate of expansion and the relaxation rate form two critical opposing factors. If the expansion rate of the volume is slow enough, then the system has sufficient time to reach maximum entropy, to achieve thermal equilibrium again. If that is not the case, the maximum entropy increases faster than the actual momentary entropy of the system. It therefore deviates more and more from equilibrium; it develops more and more order and structure. We should emphasize that all this is fully in accord with the second law of thermodynamics: the entropy is indeed continuously increasing, but not fast enough to reach the maximum value for the given volume. The resulting discrepancy means order. Requiring an increase of entropy thus does not always mean an increase of disorder, a reduction of structure, as sometimes claimed.

We should further underline that these considerations are schematic and hence certain important details are neglected. The expansion reduces the density in the given volume, which in turn generally increases the relaxation time; as a result, the actual entropy grows slower, and order and structure formation faster—at least, if the expansion rate is kept constant.

The cosmic result of our considerations is quite evident: starting from an extremely hot gas in equilibrium at initial state, the expansion leads in the long run to a discrepancy between actual and maximal entropy and thus creates structure and order in the medium. In this sense, complete disorder means that the system is in a macrostate, for which all microstates are equally likely and whose entropy is thus at a maximum. In the case of order, some of the microstates are completely excluded, and some have a higher probability than others. In terms of entropy, order or structure R is thus defined by the relation

$$R = S_{max} - S.$$

The order of the system vanishes if the entropy reaches its maximum possible value. Then all states are equally probable—no particular structure is selected. If R is not zero, we have some kind of order or structure. What its specific form is depends on the nature of the interaction between the constituents of the medium, in particular its density and temperature. It may seem surprising that disorder is more generally defined than order— that the latter can be defined only once you know the former. But order can appear in many forms, to larger or smaller extent, as rain, hail, or snow, whereas disorder just means that all microstructures are treated equally.

At this point, we also note that if we carry out the ultimate Joule experiment and simply let the initially confined gas escape into empty space, then the volume available after removing the confining wall becomes infinite, and so does the maximum possible entropy. The actual entropy of the expanding gas increases with time, but it always remains finite. The order as defined earlier can thus never vanish: the system can never achieve a maximum entropy; it can never reach thermal equilibrium.

That brings us to the second aspect mentioned earlier: how does the form of the interaction between the constituents affect the state of maximum entropy? We begin by noting that for gases and similar rather weakly interacting media, volume is the decisive factor. If we perform a Joule experiment and simply remove all walls of the container, letting the gas expand into empty space, then everything flies apart: now there is no maximum entropy—that would be infinite, just like the volume now available. In such gas-like cases, the value of maximum entropy is thus defined through the size of the system, determined "from the outside." What happens if we allow much stronger interactions?

Here a classical example is gravity, which has two extraordinary properties: its range is infinite, and it is always attractive. Electromagnetic force is in principle also of infinite range, but there are positive and negative charges, with equal charges repelling, and unequal charges attracting each other. That means a cluster containing equal numbers of positive and negative charges appears to be neutral, uncharged, if observed from far away. The different charges compensate each other's effects, so that the whole cluster does not create an electric force. In contrast, a cluster containing n particles of mass m has a total mass $M = nm$, and this total mass determines the force of gravity exerted by the cluster—there is screening of charges, whereas gravity is additive. For that reason, gravitational effects of distant bodies can never be completely neglected. And whereas a cluster of equal electric charges tends to be driven apart, a cluster of masses is always pulled together, contracted. So, in a way, gravity itself determines the size of the system, and the external "box" becomes less unimportant.

Up to now, we have considered energy and volume of the system as two independent parameters, both being determined from the outside by the experimenter. Such a picture now finds its limits. According to Albert Einstein's *general theory of relativity*, the theory addressing the role of gravitation, the presence of mass deforms, curves the space in which it is contained; more generally, that holds for the presence of energy as well. This means that an equidistribution of many masses in space no longer needs to be the state of maximum entropy. Imagine a number of marbles, equally distributed in a plane, here on Earth. If we now create a dip in the center of the plane, the marbles will roll into that hollow. In the plane, an equidistribution provided the state of maximum entropy, but in the presence of a dip, a pileup at the deepest point constitutes such a state. If we start from a set of masses initially distributed uniformly in space, the effect of gravity will have them contracting to a cluster much smaller than the starting size of the box. The size of the pile of marbles, the density of packing, is determined by the size of the individual marbles. The same holds true for the "gravitating" gas, for which the size of the individual masses determines the size of the star or the cluster of stars. In addition, there are of course effects due to motion of the constituents.

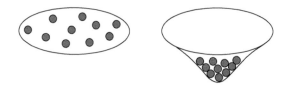

The effect of space curvature on the distribution of masses

The structure of the universe on a cosmic level is very largely a consequence of gravity. And in this case, the hot uniform gas, just after the Big Bang, is, in the long run, after expansion, no longer in the state of highest entropy. Such a state is achieved through

Cluster formation,

so that it is precisely the second law of thermodynamics that prevents the classical heat death of the universe. In other words, the evolution of the universe proceeds indeed from lower to higher entropy; but for a medium with gravitation, the uniform gas is a state of lower, and the cluster pattern a state of higher entropy. That constitutes the crucial difference from the short-range gases of the usual thermodynamics.

In our context, we can then picture the start as a system of total mass M (i.e., of total energy Mc^2), consisting of individual masses interacting through gravity. According to Einstein's general theory of relativity, such a system, even if it begins as a uniform gas in a box V, will soon determine its own fate and contract to a cluster of much smaller size. This size can be calculated, and it is much smaller than the starting volume. Up to now, we have considered energy and size of the system as two separate entities. In the case of gravity, that ends at some point—if the total mass becomes large enough, it determines the spatial extension of the system itself. And in this case, the state of maximum entropy is a distribution of clusters, a world of stars, no matter what the initial size was. In Appendix 2, we shall consider the underlying physics in a little more detail. We shall see in particular that in the presence of gravitation, an equidistributed set of masses will converge to something like a solar system in order to maximize its entropy. The masses will then circulate around a center in a way to compensate attractive gravitational force by repulsive centrifugal force.

We should note that this does not imply that the entropy of a hot gas is lower than that of the star into which it eventually collapses; the entropy of the star is in fact lower. But in the formation of the star, during the collapse, part of the total available mass is converted into radiation, photon emission. And the entropy of that photon cloud is in fact much higher than that of the initial hot gas, because the density of the cloud is much lower and hence the volume of the cloud is larger. So the entropy-increasing transition is one from a uniform gas of matter particles to a compact dense star immersed in a very much larger cloud of photons.

The general framework for the development of the universe after the Big Bang is thus provided: expansion can cause a world with structure, not in equilibrium, and even in equilibrium, long-range interactions such as gravity make a distribution of stellar clusters the state of highest entropy.

Some 500 million years after the Big Bang, we thus find a world in which the tiny irregularities created from earlier quantum fluctuations have led to cluster formation. Clouds of gas float through a now relatively empty space, collide and unite to form larger clouds or break up into smaller ones. That leads to some important features. When two clouds collide, it generally causes a rotation of the connected system, so it is not surprising that most galaxies today have a disk-like structure, with a higher density in the center.

Formation of a galaxy in the collision of two gas clouds

The dynamics of galaxy formation remains not fully understood even today. Often spiral-like arms directed outwards form at the edge of the central disk. If within a galaxy two gas clusters collide, the impact region provides a zone of extreme energy, high enough to result in nuclear fusion, which in turn leads to radiation. In this way, shining stars appear, and each galaxy consists of innumerable stars. The dimensions of galaxies surpass all imagination: in the case of our own, the Milky Way, the average thickness is about 3000 light years, with a diameter of some 100,000 light years; it contains more than 100 billion stars. And in our visible universe there are billions of such galaxies, which again combine to form supergalaxies. Again, the dominating pattern seems to be self-similar, with structures of the same kind at all scales.

Spiral galaxy Messier 101, Photo courtesy of ESA and NASA.

Perhaps the picture of Messier 101 can convey a little of the beauty and the infinity of such structures. Our home galaxy, the Milky Way, is also of this type, and our sun, the center of our solar system, is in one of the spiral arms, near the edge of the actual center.

We thus find some similarity between a galaxy and a solar system such as ours. In a galaxy, stars instead of planets circle around a center, which however is not determined by one single object. In our solar system, the mass of the sun is a thousand times larger than the masses of all planets combined, so it makes sense to speak about planets circling around the sun. In the case of the galaxy, the center is—from the point of view of our sun—determined by all stellar objects within a sphere drawn by the solar trajectory. That includes not only all stars in that region, but also an extraordinarily massive black hole at the galactic center; its mass alone is about 1% of the total Milky Way mass.

The sun then encircles the center of the Milky Way with a velocity of about 200 km/s; that means it needs some 200 million years for a complete trajectory. It is estimated that the sun is some five billion years old, so that it has made the complete orbit about 25 times. And our Earth takes part in this voyage. We have already noted one effect of this motion: because the Earth moves with such a speed relative to the center of the Milky Way, the cosmic microwave background, determined by all space, is correspondingly Doppler-shifted when seen from our spaceship Earth. The picture on page 81 shows the direction we are traveling through space.

We mentioned the size of galaxies. If we can estimate the number of stars in a galaxy and the mass of a typical star, then we know the mass of the total galaxy. The number of stars is obtained by dividing the light output of the entire galaxy by that of a typical star. From that we conclude that the Milky Way consists of about 100 billion stars and hence must have a mass of about 100 billion solar masses. Using that information, the law of gravity tells us how long it must take any given star for one complete trip around the Milky Way. Unfortunately, the result turns out to be very wrong—as we shall see in the next chapter. Although "seeing" is perhaps not the right word here, there is more in the universe than what we can see.

7

Dark Corners

There was a Door to which I found no key,
There was a Veil past which I could not see.
THE *RUBÁYIÁT* OF OMAR KHAYYÁM
(TRANSL. EDWARD FITZGERALD)

Dark corners have always been around in physics, and most likely always will be. But often the key to understanding is not found under the bright light of the lamp, so we have to search in the unlit dark part of the world. Dark corners are thus on the whole quite fruitful: they are nature's way of telling us to keep looking and to keep thinking.

Quite often the continuing search takes a very specific form. When Rutherford proposed a heavy positive nucleus surrounded by light negative electrons as the model of the atom, the total mass of the protons did not nearly suffice to give the mass of the nucleus. Something was wrong. The way out, Rutherford proposed, was to introduce new, heavy but neutral constituents that, together with the protons, would provide the measured mass of the nucleus. The new constituents simply could not be "seen." Twenty years later, his student Chadwick identified these *neutrons* experimentally.

The neutron itself continued in this vein. In an isolated state it was not stable, but decayed into a proton and an electron, keeping the overall charge zero. The masses and the kinetic energies of proton and electron could be measured, and when they were added, the sum was less than the mass of the neutron. Again something was missing. This time it was Wolfgang Pauli who blamed the deficit on another invisible particle, the neutrino. The name—little neutron—was invented by Enrico Fermi, and some twenty years after Pauli's proposal, it was indeed experimentally identified.

Ernest Rutherford (1871–1937)

Looking at the universe, we encounter even now several such dark corners, which must hide something that cannot be seen, but established only through its effect on the surrounding world. One such phenomenon we have already mentioned:

Black holes.

From our point of view, a black hole is basically what remains of a sufficiently massive star when it is completely burned out. A star acquires its size from the balance of gravitational attraction, compressing the mass, and the opposing pressure of the heat created by nuclear fusion. When there is no more nuclear fuel, the stellar fire is extinguished and the heat pressure that had compensated the gravitational force of attraction is no longer available. The star now collapses, undergoing several different reaction processes, until it consist only of neutrons. These are, as we mentioned, territorial particles that don't like to be compressed—the resulting Pauli pressure limits the possible shrinkage and leads to the formation of a much smaller stellar object, a neutron star. That is what happens provided the overall stellar mass is not too large (less than some five solar masses), because otherwise gravity is able to overcome even the Pauli pressure, compressing the neutron star further and further and thereby creating a completely new object: a black hole.

The essential property of such objects is that their force of gravity is sufficient to prevent even light from escaping their domain of attraction. On the surface of the Earth, we have the so-called escape velocity, the velocity we have to give a bullet so that it can over-come the attraction of the Earth and escape into outer space. It is proportional to the mass of the Earth and inversely proportional to the radius of the Earth. We can therefore imagine a stellar body suffi-ciently heavy and sufficiently small that the escape velocity becomes greater than the speed of light; then even light can no longer get away. Such an object retains simply everything: it is a black hole from which nothing can ever escape.

More than two hundred years ago, an English priest and natu-ral scientist, John Michell, predicted that such objects should exist, using arguments similar to the ones we just presented. The famous French mathematician Pierre-Simon de Laplace provided a math-ematical proof not much later. Today we are sure that such black holes indeed exist in our universe. One can never see them, but it is possible to observe their effect on the surroundings: if somewhere a shining beam of light suddenly disappears, a black hole has eaten it. The well-known English mathematician and writer Lewis Carroll, immortal through his *Alice in Wonderland*, has given a beautiful pres-entation of such a phenomenon in his epic poem *The Hunting of the Snark*. An expedition sets out to find the mysterious creature Snark, of which there are two species, common Snarks and Boojums. The latter have a horrible property: whoever sees one, immediately dis-appears, dissolved into thin air. How can they then be discovered? To find the solution to this dilemma—and not only for that reason—it is really worthwhile to read the story.

Things like that indeed exist in the universe. There are twin stars, orbiting around each other. If one dies and becomes a black hole, the material of its partner, still shining, is sucked into the black hole accompanying it. The twin star system Cygnus X-1 seems to be of this nature.

The black holes created out of dying stars are very massive, some 5–10 solar masses or more, but also very small: for a black hole of ten solar masses one obtains a radius of about 30 kilometers, while a shin-ing star of that mass has a radius of more than a million kilometers.

Schematic view of the Cygnus X-1: on the right is the black hole, on the left its still shining companion being sucked in. Photo courtesy of ESA and NASA

We believe today that besides stellar black holes, there is yet another form of such entities: the so-called supermassive black holes, with masses up to a billion solar masses. These monsters are found in the centers of most galaxies; they could have formed in the early stages of the universe, when gas clouds first contracted to very massive, star-like structures. Collapse would then lead to intermediate-sized black holes, which gobbled up all stars around them to reach their present size. We are quite sure today that such a monster galaxy resides in the center of our Milky Way. We will shortly see how such a conclusion can be drawn.

One of the greatest discoveries in physics is Isaac Newton's conclusion that the same force, the force of gravity, determines in a universal way the attraction between all masses. That holds from falling apples on Earth to planets orbiting the sun and beyond, and for the motion of galaxies.

The strength of the force of gravity is specified by a universal constant G, Newton's constant, that determines how strongly two given masses separated by a given distance attract each other. If we know the radius of the Earth, we can use Newton's law of gravity to "weigh" the Earth, that is, determine its mass—details are given in Appendix 2. Applied to the moon, Newton could use his law to calculate that in balance between earthly attraction and centrifugal force, it would need 28 days to orbit once around the Earth. Newton's law implies (see again Appendix 2) that the square of the orbital velocity is proportional to the distance between Earth and the moon. The weight of the moon is irrelevant for this—any satellite positioned on the lunar orbit would also need 28 days for the trip once around.

The same law can be used to calculate the orbit of the Earth around the sun, or the orbit of any other planet. In this way one obtains a rule the German astronomer Johannes Kepler had already derived before Newton: the square of the orbital velocity of a planet around the sun is inversely proportional to its distance from the sun. We can use Kepler's rule either to determine how long a year is, or—if we know that already—to "weigh" the sun.

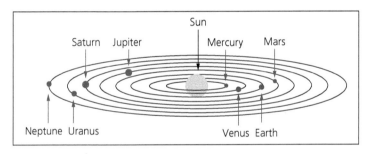

Planets orbiting the sun

And we get a universal curve connecting the orbital velocity of any planet to its distance from the sun. For the determination of this curve the mass of the planet is unimportant. To draw the curve we only have to know Newton's famous G and the mass of the sun. The weighing process just mentioned led to 2×10^{30} kg, and the corresponding curve is shown in the following picture, labeled Kepler's rule. We can now get from the astronomers the orbital velocities of the different planets and their distances from the sun and enter them in our picture. It is evident that all planets in our solar system fit extremely well into the scheme of Kepler and Newton. Even Pluto fits, although it has just lost its status as full planet and is now a labeled dwarf planet.

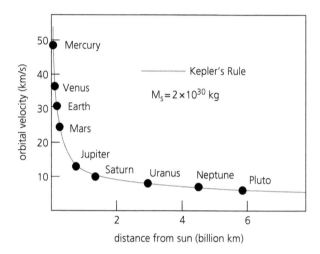

Orbital velocities versus distance from the sun for the planets of our solar system, compared to Kepler's rule

Given this information, we can now come back to the giant black hole at the center of the Milky Way. One observes a star there (S2 in the astronomical notation) that orbits an invisible center of our galaxy at a distance of about 10^{10} km every 15 years. That implies that the responsible gravitational center must have a mass of more than a million solar masses. The observed orbit requires that this immense mass must be contained in a volume of about the size of our solar system: over a million solar masses in a volume of our planetary world. No known astronomical structure, other than a black hole, can lead to such a large mass in such a small volume.

Black holes thus become for us a first instance of a dark, unattainable corner, whose darkness we can never illuminate. We can see what effect they have on the world around them, but we can never find out what happens in their interiors. Nevertheless, in comparison to what's coming next, they are still in a way quite familiar.

They are a result of the best-known force of our world, gravity, and since Michell and Laplace we can imagine how they can be formed, as a consequence of the force of gravity acting on matter as we know it.

Such considerations reach their limit, however, if we want to understand the large-scale behavior of galaxies. As already noted several times, that leads to

Dark matter.

In the following, we want to look at the reason for its appearance in more detail; but for better understanding, it seems helpful to start with the conclusion. A galaxy consists of millions or even billions of stars, which give the system a certain mass; a small part (less than 1%) is provided by the giant black hole in the center, the rest by the multitude of stars. The mass of the galaxy determines the motion of its stars or groups of stars, as specified by the law of gravitation: the attraction of the mass must compensate precisely the centrifugal force on the orbiting stars, so that the whole system is kept together. And the mass of all visible stars in our galaxy is not nearly sufficient to explain the motion of stars at the edge of the galaxy. As early as 1933, the Swiss astrophysicist Fritz Zwicki concluded that every galaxy, including ours, the Milky Way, must contain five to ten times more invisible mass, to account for the behavior of the stars at the edges: there must be very much invisible *dark matter*. It interacts with the rest of the world only through gravity; for all other forces, it does not exist. Just as with the nucleus and then with the neutron and its decay, we again encounter a situation in which we see that something is missing; there must be more than we can see, but we don't know what. Let's look at this problem now in a little more detail.

In the interior of a galaxy, stars circle around a center, which also contains the giant black hole mentioned earlier. The farther we move away from the center, the more stars are contained in the sphere defined by our position, and therefore the bigger is the mass that determines the orbit of a star on the surface of this sphere, and correspondingly the greater is its orbital velocity. Kepler's rule gave us the orbits of planets around the sun, whose mass is fixed. As a result, the orbital velocity decreased with increasing distance from the sun. But now, in the case of a galaxy, the relevant mass increases with distance from the center, proportional to the enclosed volume: the effective "sun" becomes more and more massive. As a result, the orbital velocity now *increases* with the distance from the center.

This increase stops when we reach the edge of the galaxy. The enclosed mass now no longer grows, so that Kepler's rule in its original form becomes operative: with increasing distance, the mass remains essentially constant, and so the orbital velocity for more distant, outlying stars should decrease, just as it does in our solar system. For such outlying stars, the bulk of the galaxy is something like a sun, which keeps objects outside the densely populated interior in orbit through its force of gravity. We can determine the effective galaxy mass through the mass of the visible stars, and then we can predict the orbital velocity of an outlying star in terms of its distance from the center. The situation is thus completely analogous to the case of the sun and the planets, with the bulk galaxy as sun and the outlying stars as planets.

The resulting prediction is shown in the following picture. Comparing that to the measured orbital velocities of outlying stars, we conclude that the prediction is completely wrong: the orbital velocities of the outlying stars remain constant, independent of their distance from the center of the galaxy. Planets close to the sun orbit faster than those far from the sun; but here all outlying stars, up to distances of five to ten times the size of the galaxy, all have the same orbital velocity. How is that possible?

If the orbital velocity does not decrease with increasing distance from the center, then apparently the force of gravity involved must increase. However, the visible mass remains constant, and in any case it is too small. So there must be an additional, invisible form of matter in which the entire system is embedded. And the total mass of this dark matter must increase linearly with separation from

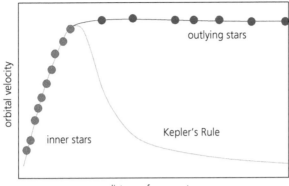

The orbital velocity of stars in a galaxy as function of their distance from the center, as measured and as predicted by Kepler's rule

the center; that implies that its density decreases with the square of the separation distance (see Appendix 2). For each outlying star, the mass responsible for its gravitational attraction is then given by the shining mass of the galaxy and in addition by the dark mass contained in a sphere of the size of its orbit. And that means the visible, shining component of a galaxy contributes only in small part to its total mass—five to ten times more remains invisible. Each galaxy we see in the sky is embedded in a much larger cloud of dark matter.

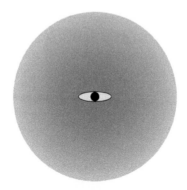

The cloud of dark matter around a galaxy

So today we can picture a galaxy as consisting of three components. In the center, there is a supermassive black hole, around
which with increasing distance an ever-growing number of stars is
orbiting. These orbits define a sort of disk, like a throwing discus,
just as the planets in their orbits define a kind of plane around the
sun, with some fluctuations. This galactic discus in turn is embedded in a sphere of dark matter, which assures that the outlying stars
circle around the center with constant orbital velocities.

While the black holes still are due to gravity effects on normal
matter—that is, effects on the compression of neutrons—dark matter brings us to the limits of our wisdom. What is it made of? The
search for an answer to this question is today probably one of the
central topics of particle physics as well as of cosmology. In other
words: we don't yet know.

We can, however, determine with measurements of increasing
precision that dark matter does not consist of any known species of
particles. It neither emits nor absorbs light, nor does it show any kind
of interaction with our visible world, except through gravity. It surrounds each galaxy like a huge cloud, whose density, as mentioned,
decreases with the square of the distance from the galactic center.
The most popular speculation today is that this cloud consists of
unknown, weakly interacting and very massive particles (Weakly
Interacting Massive Particles = WIMPs). Millions if not billions of
such WIMPs would have to be contained in each cubic meter of our
present world—but because of their extremely weak interaction,
they remain (so far) not detectable. The experimentalists at CERN
in Geneva try to produce theses ghosts in high-energy collisions,
and the theorists try to find suitable candidates in their supersymmetric theories. But, as mentioned, so far they are all still searching.

The experimental search is once more based on the scheme that
led to the discovery of neutron and neutrino: something is missing
in the observed picture. This approach was perhaps first illustrated in
one of the stories of Arthur Conan Doyle, which deals with the theft
of a race horse and the murder of its trainer. Sherlock Holmes points
out "the curious incident of the dog in the night-time." When the

accompanying Scotland Yard detective replies that the dog did nothing in the night-time, that it did not bark a single time, Holmes notes that just that was the curious incident. Something like that must happen with the WIMPs: if a WIMP is really produced in a collision, it will never be seen in the detectors, due to its extremely weak interaction. But the production of a very massive energetic particle leads to a strong recoil, and that in turn produces a jet of many normal particles, as shown in the following picture. So we have to look for a jet of many particles, whose recoil partner is missing; we have to find such unbalanced "monojet" configurations.

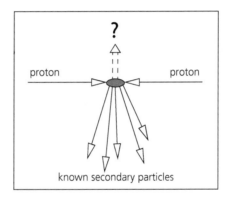

Something must balance the recoil of the observed known secondary particles: a WIMP?

Since the WIMP mass is expected to reach up to 1000 proton masses, such a search requires very high collision energies; perhaps those of the large hadron collider in Geneva are sufficient. The WIMP search is in fact one of the most pressing topics at CERN, and a success would be a dramatic breakthrough. But, unfortunately, there are other, more mundane possibilities that could account for the lack of a visible recoil—witness neutrinos.

Theorists hope, as we have indicated, that among the many particle species predicted by supersymmetric models, one might find some particles that could be the constituents of dark matter. The mathematical structure of these theories fascinates many theorists, but mathematical beauty alone is not enough, as we well know. The cycles and epicycles of the Ptolymean geocentric picture of the solar system provided a complex, mathematically very interesting scheme; moreover, it made predictions that were extremely well confirmed experimentally. Nevertheless, it was eventually replaced by the much simpler Copernican formulation. So we can only wait and hope.

Besides black holes and dark matter, there is yet another feature that in spite of much research has remained even darker and more mysterious:

Dark energy.

It appeared already at the beginning of our considerations, as the space energy remaining in our universe after the bubble in the multiverse burst. Albert Einstein introduced such a quantity 100 years ago in his general theory of relativity, with quite similar arguments. It seems worthwhile to look a little closer.

Newton's mechanics describes the force of gravity between two masses, but it does not say how and why this force comes into action. Einstein therefore went one step further: he explained the effect of the force as a consequence of the structure of space. The Earth circles around the sun, according to Newton, because the force of gravity of the sun just balances the centrifugal force experienced by the Earth in its motion, just as we can tie a stone to a string and make it circle around us. Einstein tried to understand the nature of the string and came to the conclusion that such a string does not exist. Instead the presence of the sun deforms space, just as a heavy ball on a soft surface causes an indentation, and the Earth then rolls in this bowl, circling around the sun—see the following picture.

These considerations led Einstein to his famous equations relating space, time, and matter. In simplified form (Appendix 3 contains a few more mathematical details), one can summarize these equations as

$$G_E = T_M,$$

The orbit of the Earth around the Sun as caused by space deformation

where G_E is a mathematical quantity (the Einstein tensor) describing the deformation of space, and T_M specifies the contents of space (the density of energy and matter as well as the resulting pressure). Since space and time lead to several components (x,y,z;t), the previous expression is an abbreviated form of several equations. The American physicist John Wheeler, who also coined the term "black hole," summarized the statement of these equations as

Matter tells space how to curve and space tells matter how to move.

An essential quantity determined by these equations is the scale for measuring the separation of distant objects, like galaxies: the so-called scale factor $a(t)$, which can vary with time if the universe expands or contracts. And the solution of the Einstein equations indeed led to a world which in general was not *static*. The scale factor could increase or decrease with time, and its rate of change could in turn also change with time. However, the Einstein equations in their original form predicted a continuous decrease of this rate.

Before Hubble's discovery, the idea of an expanding universe did not fit into the accepted view of the cosmos: the universe was considered as eternal and unchangeable. The equations of the form given above did not allow this; they led to an $a(t)$ increasing or decreasing with time, that is, to a finite rate of expansion or contraction

Albert Einstein (1879–1955)

$v(t) = \Delta a(t)/\Delta t$. Moreover, they implied that even this rate would decrease with time $b(t) = \Delta v(t)/\Delta t \neq 0$; nothing remained eternal and unchangeable.

To alleviate this problem, Einstein added a further term to his equations, the *cosmological constant Λ*:

$$G_E = T_M + \Lambda.$$

It was intended to compensate exactly the effects responsible for contraction or expansion, such as the attraction of gravity, and in this way allow an eternally constant universe. The problem this led to, however, quickly became evident: Λ was supposed to be a universal constant, while all other terms in the equations varied with time. That implied that the desired compensation could take place only at one specific time. If the energy density in the universe should subsequently change a bit—for example, due to galaxy formation—the compensation would be destroyed and everything would move again. For more details, we again refer to Appendix 3.

On the left side of the last equation we have Einstein's curvature tensor, which is determined by the effects on the right side, the presence of energy and matter in space. The addition of the cosmological constant implies that not only the matter and radiation contained in space play a role in this, but that space itself contains a universal "vacuum energy," constant in space and time. It only affects the structure and the evolution of space and has no further consequences. In our present view, this dark space energy is just what is provided by Einstein's cosmological constant.

Even the dark energy cannot, as we have mentioned, make the universe static, but it can in fact decide what will happen. The present frame is determined by two measurements. The precision data of the cosmic background radiation indicate, as we saw in Chapter 5, that space itself, the stage for all happenings, is intrinsically flat. What curvature there is must have arisen from the contents of space. Moreover, the investigations of supernova explosions carried out some 20 years ago have shown that the universe not only expands, but that its rate of expansion is in fact increasing—the expansion is accelerating. Such an acceleration can be achieved with a suitable Λ; for details, see again Appendix 3. The Nobel Prize in 2011 was awarded to the scientists responsible for this discovery, Saul Perlmutter, Adam Riess, and Brian Schmidt.

The presently ever-growing expansion of the universe thus requires the presence of the dark space energy, which uniformly permeates the entire space and whose density remains constant in space and, as far as we know today, in time. This density is extremely small—we shall see shortly how small. Shortly after the Big Bang, the effective volume of the universe was still small enough that gravity could overcome the expansion force of the dark energy. The universe expanded, but its rate of expansion decreased. At some point, however—cosmologists think some 500 million years ago— the volume had become so large that gravity could no longer keep up. Now there was so much space and hence also space energy that the expansion rate began to increase again, and it continues to do so today. Because the amount of matter in the universe remained constant, it constitutes today only about a quarter of the total energy of the universe—the remaining three-quarters consists of structure-less space energy, as we showed in Chapter 5.

The analysis of the measurements mentioned previously shows that the dark space energy has a value of about

$$10^{-44}\,\text{GeV/fm}^3 = 10\,\text{GeV/m}^3.$$

In terms of the units used, the radius of a proton is about one femtometer (fm) and its mass one giga electron volt (GeV). The vacuum energy of the universe thus corresponds to roughly ten protons per cubic meter, or to a density that is by a factor 10^{-44} less than that inside a proton. In yet other words: an empty space of the size of the Earth contains as much vacuum energy as a gram of water. For this reason, the dark energy becomes relevant only for dimensions of intergalactic size. But then it becomes essential.

We're already stumbling in the dark, as far as dark matter is concerned. For dark energy, things become even worse. If it would simply be zero, we could just say that empty space is empty after all.

But to understand or explain this specific small value different from zero—that is today perhaps the most serious issue for cosmology.

There are various ways to arrive at an energy for the vacuum. In Chapter 5 we saw that the physical vacuum entered the scene when the quarks combined to build hadrons and thereby allowed the existence of empty space. Before this time, all space was densely filled with colored quarks and gluons. In the simplest models, the mass of a hadron is a combination of the kinetic energy of its quarks and an inherent energy of the empty hadron "shell" or bag: this is the energy of a vacuum bubble of hadronic size. Hadron spectroscopy leads to an estimate of about 0.2 GeV/fm^3 for the energy density of this bubble, so that about four-fifths of the mass of a proton is given by the kinetic energy of its three quarks and the remaining fifth by the vacuum energy of the empty bag. This gives us an idea of what the energy density of the vacuum could be: 0.2 GeV/fm^3 look at the value determined from the expansion of the universe shows us that the hadron bag value is too large by a factor of about 10^{33}. Experts in astro-particle physics have extended these estimates to earlier stages, and find that the vacuum energy due to quantum fluctuations at the Planck scale is too large by a factor of 10^{120}. With some pride they conclude that this is the largest discrepancy between theory and experiment ever encountered in physics.

We can thus justly say that more than three-quarters of whatever is contained in our universe consists of unknown constituents. Some 5% of the contents are known particles of the world we understand; some further 20% is dark matter, perhaps somewhat similar particles we don't yet know. And three-quarters finally consists of dark space energy, of which we don't really understand anything, except that it drives the expansion of the universe. Neither its nature nor its strength can be determined in the framework of our present theories. This clearly leaves more than enough work for future generations of physicists and cosmologists. And the conclusion that more than three-quarters of the universe is of a completely unknown form constitutes yet another form of Copernican revolution. What we know and understand makes up only an extremely small fraction of all there is.

8

The End of Time

Yes, God murmured, it was a good play,
I will have it performed again.

BERTRAND RUSSELL, *A FREE MAN'S WORSHIP*

The end of time is, from a human point of view, as difficult to imagine as its beginning. Time as such is really an essential part of our existence; we can think of different spatial worlds, but in all of them time always seems to flow in the same way. At the beginning of this book, we noted that with the Big Bang both space and time appeared. Yet even if relativity theory combines the two into a space-time, essential differences remain. In space we can move in all directions, but not in time. As long as we consider dynamical theories in a formal world of one time and three space dimensions, we can move forward as well as backward in both space and time. But as soon as we include the collective effects of the real world, when we bring the second law of thermodynamics into the game, all that is over. Time acquires a direction. We get older, not younger. We may know what happened in the past, but not what may happen in the future. We can still influence events in the future, but not those in the past. The past is a subject for historians, the future one for prophets.

If the order of cause and effects is no longer valid, the world becomes a strange place. The White Queen in the second part of Lewis Carroll's *Alice in Wonderland* can remember the future as well as the past. She utters a scream because she knows that the next moment she is going to prick herself with a needle as she puts on her scarf.

In quantum physics, such things exist: an effect of anticipation as well as one of cause. An electron hit by a photon emits a further photon as the result of being hit. But for a complete description of the process one needs a further contribution, in which the electron emits the second photon before being hit by the first, in anticipation of the coming interaction. And for the two-body process of electron and photon, both the "causal" and the "acausal" contribution are crucial.

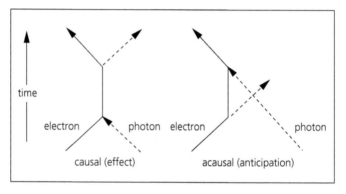

The symmetry of cause and effect in quantum physics

The direction of time appears only as an emergent effect in the macroscopic world of many particles, of many degrees of freedom. Only then does the second law of thermodynamics take hold. The gas in Joule's experiment never returns to its initial volume, even though the trajectories of two atoms in the box can go forward as well as backward. For them, there is not yet a direction of time—that appears only as the joint effort of many particles.

So there exists an order of events, and time marks subsequent points along this direction. But one can also imagine other quantities fulfilling that purpose in the chronology of the universe, for example, temperature. The primeval world started in an extremely hot stage, and the cooling connected to the expansion also determines specific points of evolution. As the result of the binding of quarks into hadrons, the physical vacuum appeared some ten microseconds after the Big Bang, or at about 10^{12} degrees Kelvin. The binding of nucleons to form atoms occurred some 380,000 years later, at a temperature of about 3000 degrees Kelvin. If we label the axis of happenings by the time, our scale runs from zero (the Big Bang) to—well, to where, to eternity? And if instead we choose temperature, the scale starts around infinity and ends, we guess, around zero. But while the time seems to flow smoothly, the temperature could in principle fluctuate. We can try to use the observable evolution to determine tendencies that we then "extrapolate into the future." But as the well-known Austro-American physicist Viktor Weiskopf once said, "Predictions are a tricky thing, particularly if they involve the future." Weiskopf knew what he was talking about, since he was one of the first directors of the European Council for Nuclear Research (CERN) in Geneva, Switzerland, when that center was just starting. So what is in store for our universe?

To investigate this question (to solve it would be too optimistic), we have to look a little more closely at the temporal evolution of the universe and of the mechanisms causing its expansion. The expansion is caused by the remaining dark space energy; gravity provides an opposing effect. Shortly after the Big Bang, the volume of our future universe was so small that gravity could dominate the expansion force of the dark energy present at that stage: the rate of expansion decreased with time. Note that it is the *rate* that decreased, not the expansion as such. If in a certain year the volume tripled, then in the next year it only doubled.

At some time, however, the volume had become so large that dark energy took over. The slowing down of the expansion stopped, and from that point on the opposite behavior started: if in a given year the volume doubled, then in the next it tripled. This switch occurred around a hundred years after the Big Bang, and the new pattern continues until today. At the time of last scattering, less than 1% of the overall energy of the universe was dark energy, whereas today it is 75%. And if nothing happens, this development will continue.

The behavior we just sketched leaves only

Three possibilities

for the chronology of the universe. The evolution up to now and the different possible future forms are illustrated schematically in the following picture.

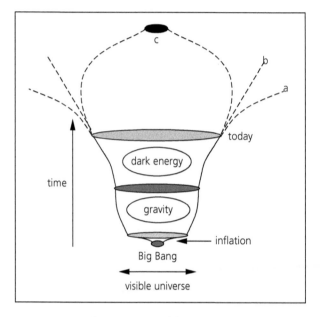

The expansion of the universe

After the termination of the Big Bang inflation, the expansion rate initially decreased, because gravity overcame the expansion effect of the dark energy: there is simply not enough space. Nevertheless, the expansion continued and as soon as space had become large enough; the amount of dark energy became sufficient to increase the expansion rate again. In this stage of ever-growing expansion rate we find ourselves (still?) today, according to the latest measurements carried out at the end of the nineties of the last century.

Future development will be decided by the dark energy. If its density persists unchanged, the expansion of the visible universe will continue undiminished (pattern a). It remains possible, however, that the strength of the dark energy will decrease just enough to create a balance between it and gravity. Then the expansion would continue a constant rate (pattern b). And finally, it is possible that the effect of the dark energy is somehow diminished, reduced in the course of time, so that in the long run gravity wins—then it would cause the world to contract again (pattern c). What consequences do these quite different future perspectives lead to?

If dark energy remains victorious in the long run, then expansion will continue, either at an increasing rate (a) or uniformly (b). That means for the future universe, there is neither a spatial nor a temporal limit. Distant galaxies become ever more distant and finally disappear beyond the Hubble horizon of what for us is within reach of sight. For us, their lights are turned off. Our own galaxy is compact enough to keep it in shape by gravity—the amount of dark energy contained within its volume is too small to tear it apart. But "our" stars eventually also stop shining, because they will have used up all their fuel. Possibly some new stars are formed, but eventually also they lose their power: we arrive at

The ultimate nightfall.

The universe becomes ever larger, but also ever colder and ever emptier. The temperature of the cosmic background radiation drops to zero, since its wavelength is dilated more and more. Finally, our world, once filled with billions of galaxies, contains nothing more but our own small world, our Milky Way, in which more and more dead stars are circling. All that was farther away has been shifted into an unattainable distance. If the strength of the dark energy remains what it is today, our galaxy is saved, because gravity is sufficient to keep it bound. If, however, the density of dark energy should *increase* sufficiently with time, then even that is no longer assured: now everything could be torn apart, even matter and atoms. In the final stage, lone nucleons and electrons would then be flying around an otherwise empty and ever-larger space. The world would then have died a *thermal death*—it would have become a uniform and ever more dilute gas. The formation of structure we described in Chapter 6 was largely based on gravity, and since its role is now over, there is no more structure. Time is not over, but it has become meaningless: there are no more events that can define a sequence. The world has once more become timeless.

In the other case, the presently still increasing expansion would start to become less, eventually stopping altogether, and finally the universe would start to contract. To have that happen, the nature of the dark energy would of course have to change; it would somehow have to deteriorate, so that it could be overcome by gravity. Then everything would contract, but this would not lead back to the original starting state; the transition from false to true ground state can never be reversed. Accidental collisions, no matter how strong, can never bring the ball back from the valley to come to rest at the top of the hill. The end is now decided by gravity. The result is a gigantic black hole, from which nothing can ever escape and which holographically stores on its surface all the information of the previous world. The amount of information determines the extent of the surface.

In this case, our space as well as our time would have been a limited show. It may be a consolation that other such bubbles appear continuously—the soup remains boiling—even though we are no longer around to explain the next universe. In a way, such a picture is esthetically quite pleasing after all; our own finite life allows us to imagine a universe that also has a finite life. And it as well is aging, from a young, hot gas to an old, cold black hole. And the finite life we witness is only that of our own universe—there are infinitely many others that appear, exist, and vanish, just as our descendants continue, live and vanish, with their descendants carrying on to make evolution endless.

At this point, we briefly mention an alternative approach that has appeared in various contexts over time. Its first appearance is perhaps the concept of *samsara*, reincarnation, an aspect of all Indian religions. It has also been pursued by ancient Greek and subsequent Western philosophers. Could it not be that also the—our—universe will be reborn in the end? Such cyclic cosmic chronologies have been attempted at various times and have for quite some time failed because of the second law of thermodynamics. More recently, scenarios of this nature have been put forward by Paul Steinhardt and collaborators in Princeton. They require that the Big Bang picture presented here, with a transition from false to true normal state and the resulting inflation, has to be given up. Instead, our universe is created in a collision of structures in a higher dimensional world, and rebounds of such collisions could produce a cyclic chronology. It would in any case also require a decreasing expansion rate of our universe, for which so far cosmologists have found no indications; but perhaps the last word has not yet been spoken.

Even if our present universe should finally end as a black hole, that does not mean the ultimate end. Only in the framework of classical physics can nothing escape from a black hole. Some fifty years ago, the well-known English astrophysicist Stephen Hawking found a quantum-mechanical way out. Empty space, the "nothing," is only on the average empty. There are always short-time fluctuations, in which a particle-antiparticle pair comes up out of the depths of nothing, enters reality, and very shortly afterward annihilates and thus disappears again. If this process happens at the surface of a black hole, there is the possibility that the black hole grabs one of the pair and draws it into its interior. That means the other particle now is "saved": it no longer has a partner to annihilate with. Energy conservation requires that someone has to pay for this process; initially there was a black hole of a certain mass, and afterwards a black hole plus the leftover particle. The mass of the black hole must have been reduced by the mass of this particle. As observed from the outside (even though there is no one left to observe), it appears as if the black hole is radiating: it emits so-called Hawking radiation. And because black holes emit more and more such radiation as their mass decreases, they will evaporate in the long run—in the very long run, however. The Big Bang was some 10^{11} years ago; estimates of a very speculative nature indicate that it will take more than 10^{100} years before our universe is finally evaporated, with a black hole as the last intermediate step. After the evaporation, only electrons, positrons, photons, and neutrinos will remain in the form of an immensely dilute ideal gas in an otherwise empty space.

So, barring reincarnation, there seems to be no escape from the second law of thermodynamics: in the end, there exists neither structure nor order nor time. In biblical words: *All are from dust, and to dust all return.* Our lives and that of our universe fit in-between.

How Many Configurations of Balls Are There?

In this appendix, we want to illustrate the counting of configurations, of microstates, using balls in boxes with compartments. Let's begin with a simple case, two balls and a box with four compartments. For the first ball, there are four possibilities, for the second there remain three: in total 12 different configurations. To obtain this result, we have assumed that the balls are distinguishable, say, one red, one blue. If that is not the case, we counted too many: we have to count configurations arising from the interchange of two balls only once. So for identical balls, we get only six configurations, as shown in Figure.

We now turn to four balls in nine compartments. For the first ball, there are nine possibilities, for the second eight, and so on, so in total we get

$$9 \times 8 \times 7 \times 6 = 3024$$

configurations. But here we also want to consider identical balls, so that we have to divide by $4 \times 3 \times 2 = 24$. That gives us 126 different configurations for four identical balls in nine compartments.

For nine balls in nine compartments, there is of course only one possibility. But doubling the number of compartments now leads to

$$\frac{18 \times 17 \times \ldots \times 10}{9 \times 8 \times \ldots \times 2} = 48,620$$

configurations. If we take the balls to be gas atoms and use as a starting point a Joule experiment in which nine balls are in a box with nine compartments, then opening the divider (doubling the number of compartments) gives with equal a priori probabilities for all configurations the chance of finding once more all balls in the "left" (nine-compartment) part of the box 1:48,620.

For the general case, we consider n balls in a box of xn compartments, where $x \geq 1$ is supposed to be a whole number. The number of possible configurations then becomes

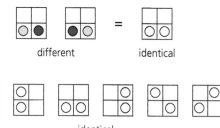

different   identical

identical

Two balls in four compartments

$$\frac{xn \times (xn-1) \times (xn-2) \times \ldots \times (xn-n)}{n \times (n-1) \times (n-2) \times \ldots \times 1} =$$

$$= x^n \left[\frac{(1-(1/xn)) \times (1-(2/xn)) \times \ldots \times (1-((n-1)/xn))}{(1-(1/n)) \times (1-(2/n)) \times \ldots \times (1-((n-1)/n))} \right].$$

For sufficiently large x and n, with $x \gg n \gg 1$, the factor in square brackets approaches unity, so that the number of possible configurations becomes x^n, as quoted in Chapter 6.

Orbits and Dark Matter

In this appendix, we want to elaborate a bit the relationship between gravity and centrifugal force, as they apply in particular to planetary orbits. In the second part, we address the role of gravity in maximizing the entropy of a system.

The law of gravity, as formulated by Isaac Newton, describes the attractive force between two masses M and m, separated by a spatial distance R. The law says

$$F = G\frac{Mm}{R^2}, \tag{1}$$

where $G = 6.7 \times 10^{-11}\, m^3/kg\, s^2$ is the universal constant of gravity. The law is a special case of the first law of mechanics,

$$F = ma, \tag{2}$$

giving the acceleration that a mass m experiences due to the action of a force F. On the surface of the Earth, we can measure the acceleration caused by gravity. As already shown by Galileo Galilei, it is the same for all masses, 9.8 m/s^2. If we know the radius of the Earth ($6.4 \times 10^6\, m$), we can use the last two equations to get

$$M = \frac{aR^2}{G} = 5.3 \times 10^{24}\, kg \tag{3}$$

for the weight of the Earth. So this is in fact a way to weigh it.

The orbits of the planets around the sun are determined by the interplay of the sun's gravity and the opposite centrifugal force,

$$K = \frac{mv^2}{R} \tag{4}$$

caused by the circular motion of the planets. Here we denote with m the mass of the planet and with R the radius of its orbit around the sun, which for simplicity we take to be circular; v denotes the velocity of the planetary motion. Newton's first big success was that with this equation, he could

determine how long the moon would take to circulate once around the
Earth. From equations (1) and (4) we obtain for its velocity

$$v^2 = \frac{GM}{R},$$
(5)

where M denotes the mass of the Earth and $R = 4 \times 10^8$ km the distance
between Earth and moon. With $2\pi R$ for the orbit, again assumed to be circular, we find

$$T^2 = \frac{(2\pi)^2 R^3}{GM},$$
(6)

for the passage time T : 30 days. And the same equation (5) determines
the orbits of the planets around the sun, if we take M to be the mass of
the sun. In this way, we obtain Kepler's rule $v^2 \sim GM/R$, which, as we saw,
is indeed followed by all planets of the solar system. For this, we need the
mass of the sun; it can be determined, once we know the passage time for
the Earth (one year) and the distance Earth–sun (1.5×10^{11} m). The result
is $M_{sun} = 2 \times 10^{30}$ kg, so that the sun is half a million times more massive
than the Earth.

If now the stars at the edge of galaxies show orbital velocities independent of their distance to the center of gravity of the system—that is, if they
remain constant for increasing R—then equation (5) necessarily leads to
the conclusion that the mass must increase linearly with R. The visible
mass, however, remains almost constant for stars beyond the edge of the
galaxy; hence we need the mentioned invisible dark matter to obtain a constant orbital passage. The density of dark matter, M_{dm}/R^3 then decreases as
R^{-2}, given that the dark matter mass increases linearly in R.

Cosmological Constant and Dark Energy

In this appendix, we want to show in a little more detail how Einstein arrived at his cosmological constant, and why today we interpret this as dark energy. For readers without any physics background, things will be a bit rough; but I hope that they will understand the conclusions even if they have to jump over the mathematical argumentation.

We begin with the original equations proposed by Einstein,

$$R_{\mu\nu} - \frac{1}{2}Rg_{\mu\nu} = \frac{8\pi G}{c^4}T_{\mu\nu}, \tag{1}$$

where G is Newton's gravitational constant and c the velocity of light. The indices μ and ν indicate space and time and hence take on the values 0 (time) and 1,2,3 (space). The left side of the equation specifies the curvature of spacetime in terms of the metric tensor $g_{\mu\nu}$. The above relation fixes the spacetime structure as determined by the energy-momentum tensor $T_{\mu\nu}$, defined in terms of the energy density ρ and the resulting pressure p. The equations thus give us $g_{\mu\nu}$ as the solution of an interplay of curvature operators and spacetime content. If the space is empty, i.e., for $T_{\mu\nu} = 0$, we have as the simplest solution the flat Minkowski space, with a diagonal matrix $g_{\mu\nu} = 1, -1, -1, -1$, i.e., there is no curvature. Other, more general solutions allow for the presence of black holes surrounded by empty space. The details are not really so necessary for an understanding of these equations; the crucial feature is that the contents of space (the right side of the equation) determine the curvature of spacetime (the left side).

One result of these equations is that due to its contents, the size of the universe can change in time. A measure of this change is the so-called scale factor $a(t)$, which fixes the scale for the determination of the size. As solution of the Einstein equations, one obtains two relations for this scale factor, relations that were first derived by the Russian physicist Alexander Friedmann. The first of them,

$$H^2 \equiv \left(\frac{\dot{a}}{a}\right)^2 = \frac{8\pi G}{3}\rho - \frac{k}{a^2}, \tag{2}$$

determines the time rate of change $v = da/dt \equiv \dot{a}$ of the scale factor. With the $H = \dot{a}/a$ for the Hubble constant H this immediately leads to Hubble's law,

$$v = \dot{a} = Ha, \tag{3}$$

according to which distant galaxies recede faster the farther away they are. In equation (2), the energy density in space is given by ρ while k specifies the form of space (see Chapter 5). For flat space we have $k = 0$, for spherical space $k = +1$, and for hyperbolic $k = -1$. As we noted in Chapter 5, the data of the cosmic microwave radiation indicate a flat space; but for the time being we'll keep all possibilities.

The second Friedmann equation determines the rate of change of the scale velocity $\ddot{a} = \dot{v}$

$$\frac{\ddot{a}}{a} = -\frac{4\pi G}{3}(\rho + 3p). \tag{4}$$

This last equation showed that Einstein's quest for a static universe was ruled out by his equations in their original form. Even if for a spherical space structure $(k = +1)$ equation (2) would only momentarily allow $\dot{a} = 0$; equation (4) shows that that soon will change again.

At this point Einstein noted that the mathematical structure of his equations allowed a modification: one could add a term on the left side

$$R_{\mu\nu} - \frac{1}{2}Rg_{\mu\nu} + \Lambda g_{\mu\nu} = \frac{8\pi G}{c^4}T_{\mu\nu}, \tag{5}$$

with Λ denoting a universal, positive quantity constant in both space and time: the *cosmological constant*. This modification resulted in correspondingly modified Friedmann equations:

$$H^2 \equiv \left(\frac{\dot{a}}{a}\right)^2 = \frac{8\pi G}{3}\rho - \frac{k}{a^2} + \frac{\Lambda}{3}, \tag{6}$$

and

$$\frac{\ddot{a}}{a} = -\frac{4\pi G}{3}(\rho + 3p) + \frac{\Lambda}{3}. \tag{7}$$

At first sight, that seemed to satisfy Einstein's wish for a static universe: for a spherical structure $(k = +1)$ there were common values of ρ, p and Λ leading to $\dot{a} = 0$ as well as to $a = 0$

But the joy did not last long. It soon became clear that any small temporal change of the energy density ρ would destroy the static nature again. Einstein's universe was effectively as stable as the ball on the hill: any small perturbation will cause it to roll down. In the same way, the static universe was not a stable situation: any density fluctuation would cause it to expand or contract.

As this point Hubble's discovery entered the scene: the universe was expanding; it was not static. Einstein noted with regret that he had missed the chance of predicting such an expansion. It is said that he called the introduction of the cosmological constant the "biggest blunder" of his life. We know today that that was not the case—as usual he was simply ahead of his time. For $\Lambda = 0$ one can indeed get an expanding universe, but the rate of expansion decreases with time. To agree with the latest supernova data, the acceleration $\ddot{a}$ has to be positive, and that requires a sufficiently large and positive Λ.

Today's new scenario of multiverse and inflation can be fitted without problem into the given framework. One only had to move Λ from one side of the equation to the other,

$$R_{\mu\nu} - \frac{1}{2}Rg_{\mu\nu} = \frac{8\pi G}{c^4}T_{\mu\nu} - \Lambda g_{\mu\nu}. \tag{8}$$

Now the curvature of space and its evolution are determined not only by the externally introduced energy density in T, but in addition by the space energy density Λ, the dark energy. And if we now measure the acceleration of the spatial expansion, as was done the supernova experiments mentioned earlier, then we can fix the value of Λ. That is how the values given in Chapter 7 were obtained.

Some references for further reading

This list is in no way exhaustive—it is meant simply to indicate some material of help for further information. It also does not aim to cite original scientific works; however, these are in general referred to in the books or surveys mentioned.

Martin Bojowald, *Once before Time: A Whole Story of the Universe*. New York: Knopf, 2010.

Brian Clegg, *Before the Big Bang: The Prehistory of Our Universe*. New York: St. Martin's Press, 2009.

Alan Guth, *The Inflationary Universe: The Quest for a New Theory of Cosmic Origins*. Reading, MA: Addison-Wesley, 1997.

Brian Greene, *The Hidden Reality: Parallel Universes and the Deep Laws of the Cosmos*. New York: Knopf, 2011.

David Layzer, *Cosmogenesis—The Growth of Order in the Universe*. New York: Oxford University Press, 1990.

Andrei Linde, The self-reproducing inflationary universe. *Scientific American Special Edition Cosmos*, Spring 1998, 9(1), 98–104.

Roger Penrose, *Cycles of Time: An Extraordinary New View of the Universe*. London: Bodley Head, 2010.

Helmut Satz, *Ultimate Horizons—Probing the Limits of the Universe*. Berlin: Springer Verlag, 2013.

Paul J. Steinhardt, The inflation debate: is the theory at the heart of modern cosmology deeply flawed? *Scientific American*, April 2011, 304(4), 36–43.

Erik Verlinde, On the origin of gravity and the laws of Newton. *Journal of High Energy Physics* April 2011, 29. This is the original reference; for a general discussion, see e.g. Dennis Overbye, A scientist takes on gravity. *The New York Times*, July 12, 2010.

Alexander Vilenkin, *Many Worlds in One: The Search for Other Universes*, New York: Hill and Wang, 2006.

Person Index

Subject Index